Southern Space Studies

The Southern Space Studies series presents analyses of space trends, market evolutions, policies, strategies and regulations, as well as the related social, economic and political challenges of space-related activities in the Global South, with a particular focus on developing countries in Africa and Latin America. Obtaining inside information from emerging space-faring countries in these regions is pivotal to establish and strengthen efficient and beneficial cooperation mechanisms in the space arena, and to gain a deeper understanding of their rapidly evolving space activities. To this end, the series provides transdisciplinary information for a fruitful development of space activities in relevant countries and cooperation with established space-faring nations. It is, therefore, a reference compilation for space activities in these areas.

The volumes of the series are peer-reviewed.

More information about this series at http://www.springer.com/series/16025

Annette Froehlich

Editor

Space Fostering Latin American Societies

Developing the Latin American Continent through Space, Part 1

 Springer

Editor
Annette Froehlich🄳
SpaceLab
University of Cape Town
Rondebosch, South Africa

ISSN 2523-3718   ISSN 2523-3726   (electronic)
Southern Space Studies
ISBN 978-3-030-38914-7   ISBN 978-3-030-38912-3   (eBook)
https://doi.org/10.1007/978-3-030-38912-3

This Springer imprint is published by the registered company Springer Nature Switzerland AG
The registered company address is: Gewerbestrasse 11, 6330 Cham, Switzerland

Foreword

Experts from the Latin American region have contributed immeasurably to the advancement of space science, technology applications, policy, and law since the beginning of the Space Age. The vast territories of Mexico, Central America, and South America underscored the need for technology to monitor the environment, human settlement, forest and farmland development, and communications among people in isolated regions. That is why the perspectives shared in this publication is so valuable. The world community is looking to the promise of the implementation of the UN Sustainability Development Goals and space technology plays a crucial role in all of them. Without our presence in space, it would be impossible to achieve the better life we are all seeking through the SDGs. The Latin American experience is instructive as we move forward, and this series will be foundational for demonstrating the value of space exploration for the betterment of the human condition. At the 2019 session of the UN General Assembly, an unprecedented event was organized by the US Department of State, the governments of Zambia and Italy and the UN Office for Outer Space Affairs entitled "Bringing the Benefits of Space to Everyone, Everywhere." The event highlighted the value of space programs and activities, and the increasing size and importance of the commercial space sector. During the event, developing countries made an impassioned case for the importance of space applications to the citizens of developing countries, a theme amplified by government and industry representatives. This event and the publication of this series during the year that we celebrate the first human to set foot on soil not of the Earth further captures the imagination of our general public on the unifying character of space cooperation and the ability of all peoples to benefit from the collective space enterprise.

Kenneth Hodgkins
Director, Office of Space
and Advanced Technology
US Department of State
Washington, DC, USA

Contents

The Need for the Enactment of a Mexican National Legislation for Space Activities

María Clarisa Jiménez Martínez

Abstract

Space activities have been exponentially increasing and have gained importance between States and private undertakings. The development of a national space law is a growing trend in order for States to regulate private sector involvement in an activity that was initially carried out exclusively by governmental agencies. However, there are several factors such as the absence of a strong space policy, that have affected the development of fruitful space activities, as is the case in Mexico. This article will provide a general overview of the current Mexican space activities under the corresponding Mexican legal framework in force, to determine if such framework promotes the development of such activities and, if required and feasible, to propose amendments to the legal framework.

1 Introduction

During the early development of space activities, the main actors were States, since governments had "the technical and financial capabilities to carry out the exploration and use of outer space."[1] Therefore, the international legal framework developed at the time focused primarily on providing for State activities in outer space.

---

[1]Tronchetti, F. (2013). *Fundamentals of Space Law and Policy*. New York, New York: Springer New York, p. 25.

M. C. Jiménez Martínez (✉)
International Institute of Air and Space Law, Leiden University, Leiden, Netherlands

© Springer Nature Switzerland AG 2020
A. Froehlich (ed.), *Space Fostering Latin American Societies*, Southern Space Studies, https://doi.org/10.1007/978-3-030-38912-3_1

1

As the years went by, the initial scenario began to change, and private undertakings started to take part in the space sector. As a result, a need for the legal framework to be adapted in order to suit both, governmental and private space activities, began to emerge.

This article intends to determine if the Mexican legal framework currently in force is enough in order to ensure the development of Mexican space activities.

In order to provide a grounded answer, the author will provide a general overview of the space activities developed by Mexico and then analyze its current legal framework, including international treaties and national law, to determine if amendments are required, and if so, under which terms.

2 Overview of the Mexican Space Activities

The satellite infrastructure is the main, if not the only, space activity developed by the Mexican State.

In 1985, Mexico launched its first satellites, Morelos 1 and Morelos 2 that composed the Morelos Satellite System, with the objective of connecting the country with domestic infrastructure. At the time, this became a historical milestone regarding the satellite infrastructure. The Morelos Satellite System set the basis for the following satellite systems developed by the country.

Telecomunicaciones de Mexico, also known as Telecomm, a decentralized public entity operated satellite telecommunications in Mexico at the time. However, in 1997, after the privatization of the satellite sector, the company Satelites Mexicanos, S.A. de C.V., referred to as Satmex, was incorporated as a private undertaking. During this period, the Solaridad System and the Satmex System were developed and widely expanded the capacities of use and exploitation of satellite communications. Later, in 2014, Satmex was acquired by the French company Eutelsat.

In 2005, the Mexican government granted the company Quetzsat, S. de R.L. de C.V. a concession to provide satellite broadcasting services, becoming the second national operator with the launch of the QuetzSat-1 satellite in September 2011.

However, in 2009, recognizing the need for a satellite system owned by the State, the Mexican Satellite System, Mexsat, was created. Mexsat is composed by the satellites Bicentenario, launched in 2012, and Morelos 3, launched in 2015, and is operated by Telecomm from control centers located in Mexico City and Hermosillo, Sonora.

The Mexican satellite activities are controlled directly by the Ministry of Communications and Transportation, without the involvement of the Mexican Space Agency. Furthermore, satellite activities are subject to Federal Law on Telecommunications and Broadcasting and the competent authority is the Federal Telecommunications Institute.

3 Development of National Space Law

Global space activities have increased at an accelerated rate, with more and more countries getting involved in their development. Private undertakings are taking part in the space market because on one hand, States needed to find new forms of financing for space activities, and on the other, technological improvements resulted in lower costs for private investors.[2]

This incursion began with cooperation between government agencies and private entities for the execution of projects that were usually developed by the public sector. Later on, private undertakings developed their own space projects.

Whereas States use space objects in order to serve public interests, including weather forecasting, research, earth observation, science, communication, and navigation; private undertakings began to use outer space mainly for commercial purposes, especially for telecommunication services,[3] global positioning systems, space launch services, and remote sensing.[4]

The evolution of space activities raised questions and challenges in relation to the application of space law, considering that "the existing body of international space law was enacted in a time frame where nation states were the main actors in space."[5] As private undertakings are now active in outer space activities, the issue is that such companies "are not directly bound by international treaties and other norms of public international law."[6] Space law arose against the typical procedure where each State legislates on a specific matter and subsequently meets with other sovereign States to try and bring their regulations closer together under an international agreement. Instead, space law first appeared in the form of international law, with international treaties and soft law. Only afterwards did the States themselves began to take into account this set of international standards when implementing and regulating this new dimension of human activity within their respective national legal systems.[7]

The five international treaties on outer space[8] intend to provide the rights and obligations of actors, whether public or private, that explore and use outer space, regardless of the nature thereof. Therefore, States that have ratified the treaties have the responsibility to ensure that private undertakings comply with such obligations.[9]

---

[2]*Supra* note 1.

[3]Marboe, I. (2015). National space law in *Handbook of Space Law*. Research Handbooks in International Law series. United Kingdom: Edward Elgar Publishing, p. 127–204.

[4]Hermida, J. (2009). Law reform and national space law: A participatory approach to space law making in developing countries. *Annals of Air and Space Law, 34*, p. 895–912.

[5]Linden, D. *The impact of national space legislation on private space undertakings: a regulatory competition between states?* September 2017, https://ghum.kuleuven.be/ggs/publications/working_papers/2017/190linden (last accessed on March 19, 2019) p. 4.

[6]*Supra* note 6, p. 127.

[7]Velázquez Elizarrarás, J. (2013). El derecho del espacio ultraterrestre en tiempos decisivos: ¿estatalidad, monopolización o universalidad? *Anuario Mexicano De Derecho Internacional, 13*, p. 583–638.

[8]See Sect. 4.1.

[9]*Supra* note 6.

This duty is expressly stated in Article VI of the Outer Space Treaty as follows:

Article VI

States Parties to the Treaty shall bear international responsibility for national activities in outer space, including the Moon and other celestial bodies, *whether such activities are carried on by governmental agencies or by non-governmental entities*, and for assuring that national activities are carried out in conformity with the provisions set forth in the present Treaty. *The activities of non-governmental entities* in outer space, including the Moon and other celestial bodies, *shall require authorization and continuing supervision by the appropriate State Party* to the Treaty... (emphasis added).

Article VI provides on how States should make sure that space activities are carried out in accordance with the Outer Space Treaty, that is by authorizing and continuously supervising private undertakings. Thus, this does not necessarily imply that States must create a national space law, but rather than States shall have the means necessary for the authorization and supervision of private space activities.[10] As such, Article VI does not set forth specific substantive requirements on authorization and continued supervision.[11] However, "Article VI of the Outer Space Treaty represents the most important legal basis for national space legislation."[12]

Even though States can guarantee the fulfillment of their authorization and continuing supervision by alternative means, the most self-evident method is enacting national space legislation.[13] "The increasing commercialisation and privatisation of space activities has fuelled [sic] the impetus for States to adopt national legislation to ensure that they are able to authorise and supervise space activities in accordance with their international obligations."[14] Within this new developing trend, national legislation was created "according to the specific needs and practical considerations of the range of space activities conducted and the level of involvement of non-governmental entities"[15] including provisions that deal with licensing and registrations processes, as well as insurances.[16]

The national space law that has been adopted by States addresses several issues and has different scopes that range from "launching of objects into outer space, the operation of a launch or re-entry site, the operation and guidance of space objects, in some cases the design and manufacturing of spacecraft, the application of space science and technology such as that used for Earth observation and

---

[10]Ibid.

[11]Froehlich, A., Seffinga, V. (2018). *National Space Legislation*, Vol. 15, Studies in Space Policy. Cham: Springer International Publishing.

[12]*Supra* note 6, p. 132.

[13]*Supra* note 13.

[14]Dempsey, P. (2013). The emergence of national space law. *Annals of Air and Space Law, 38*, p. 303.

[15]United Nations Office for Outer Space Affairs. *National Space Law*, n.d., http://www.unoosa.org/oosa/en/ourwork/spacelaw/nationalspacelaw.html (last accessed on March 19, 2019).

[16]*Supra* note 13.

telecommunications, and exploration activities and research"[17] depending on the activities that are effectively carried out by private undertakings in any given State. Typically, national space legislation is paired with national space policies usually focusing on pursuing public interest objectives such as public safety, economic growth, technological and scientific development, and national security.[18]

The United Nations Office for Outer Space Affairs (UNOOSA) offers a compilation of national space laws adopted by States. At the time of writing, 23 States have submitted to UNOOSA their national laws, with Argentina, Brazil, and Chile being the only Latin American countries that have space regulations currently in force. Unfortunately, "in the Latin American and Caribbean region, space law is, so far, almost unknown to the overwhelming majority of the public"[19] and the enactment of national space legislation is rarely seen in the region.

If Mexico wants to promote the development of space activities and become a leading actor within the space sector in the Latin American region, a national law for space activities must be adopted.

4 Mexican Legal Framework in Relation to Space Activities

4.1 International Treaties

The Committee on the Peaceful Uses of Outer Space (COPUOS), set up by the United Nations General Assembly in 1959, "is the forum for the development of international space law. The Committee has concluded five international treaties ... on space-related activities."[20]

Each of the treaties stresses the notion that outer space, the activities carried out in outer space, and whatever benefits might be accrued from outer space should be devoted to enhancing the well-being of all countries and humankind, with an emphasis on promoting international cooperation.[21]

---

[17]United Nations, General Assembly, Committee on the Peaceful Uses of Outer Space, Legal Subcommittee, Report of the Working Group on National Legislation Relevant to the Peaceful Exploration and Use of Outer Space on the work conducted under its multi-year workplan, A/AC.105/C.2/101, (April 3, 2012).
[18]*Supra* note 13, p. 318.
[19]Dos Santos, A., Filho., J. (2008). Need for a National Brazilian Centre of Space Policy and Law Studies. *Space Policy, 24,* p. 6–9.
[20]United Nations Office for Outer Space Affairs. *Space Law Treaties and Principles*, n.d., http://www.unoosa.org/oosa/en/ourwork/spacelaw/treaties.html (last accessed on March 20, 2019).
[21]Ibid.

The status of the United Mexican States regarding each of the five international treaties is as follows:

(a) The Treaty on Principles Governing the Activities of States in the Exploration and Use of Outer Space, including the Moon and Other Celestial Bodies, also known as the Outer Space Treaty, was adopted by the United Nations General Assembly, hereinafter the General Assembly, on December 19, 1966, and opened for signature on January 27, 1967. It entered into force on October 10, 1967.[22] Up to January 2018, 107 States have ratified the Outer Space Treaty and there are 23 States that have signed the treaty but have not completed the process of ratification.[23]

Mexico has signed and ratified the Outer Space Treaty, which was published in the Federal Official Gazette on November 14, 1967.[24] It is important to emphasize that the Mexican government concluded the ratification process in the same year the Outer Space Treaty was opened for signature, indicating the commitment of Mexico to cooperate at an international level in the promotion of space activities.

(b) The Agreement on the Rescue of Astronauts, the Return of Astronauts and the Return of Objects Launched into Outer Space, referred to as the Rescue and Return Agreement, was adopted by the General Assembly one year later after the adoption of the Outer Space Treaty, that is on December 19, 1967 and open for signature on April 22, 1968. The Rescue Agreement entered into force on December 3, 1968.[25] The Rescue and Return Agreement has been ratified by 96 States and 23 countries are signatories in process of ratification.[26]

The Rescue and Return Agreement has been signed and ratified by Mexico. Such commitment was published in the Federal Official Gazette for its enactment on September 20, 1969.[27] Again, the Mexican government did not even take a year to ratify the Rescue Agreement after its entry into force.

(c) The Convention on International Liability for Damage Caused by Space Objects, referred to as the Liability Convention, adopted by the General Assembly on November 29, 1971, was opened for signature on March 29, 1972 and entered into force on September 1, 1972.[28] Ninety-five countries

---

[22]United Nations, General Assembly, Committee on the Peaceful Uses of Outer Space, Legal Subcommittee, *Status of International Agreements relating to activities in outer space as at 1 January 2018*, A/AC.105/C.2/2018/CRP.3, April 9, 2018, p. 1.

[23]Ibid., p. 10.

[24]Mexican Federal Official Gazette. DECRETO por el que se aprueba el Tratado sobre los principios que han de regir las actividades de los Estados en la exploración y utilización del espacio ultraterrestre, incluso la luna y otros cuerpos celestes, abierto a la firma en Washington, Londres y Moscú, el 27 de enero de 1967, November 14, 1967.

[25]*Supra* note 24.

[26]*Supra* note 25.

[27]Mexican Federal Official Gazette. DECRETO relativo a la promulgación del Acuerdo sobre el Salvamento y la Devolución de Astronautas y la restitución de objetos lanzados al espacio ultraterrestre, September 20, 1969.

[28]*Supra* note 24, p. 2.

have ratified the Liability Convention and 19 have signed such Convention and in process of ratification.[29]

Mexico is one of the 95 States that have signed and ratified the Liability Convention, which was enacted by its publication on the Federal Official Gazette on August 8, 1974.[30]

(d) The Convention on Registration of Objects Launched into Outer Space was adopted by the General Assembly on November 12, 1974, opened for signature on January 14, 1975, and entered into force on September 15, 1976.[31] The Registration Convention has 67 ratifications and 3 signatories in the process of ratifying it.[32]

Mexico also signed and ratified the Registration Convention, which was enacted on March 23, 1977.[33] It is evident that the Registration Convention is not as successful as the Outer Space Treaty, as shown by the decreasing number of ratifications. Several issues have been identified resulting from the Registration Convention that are not to be addressed in this article; nonetheless, Mexico signed and ratified the Registration Convention.

(e) The Agreement Governing the Activities of States on the Moon and Other Celestial Bodies or the Moon Agreement was adopted on December 5, 1979 and opened for signature on December 18, 1979. It was until July 11, 1984 that the Moon Agreement entered into force.[34] Only 18 countries have ratified the Agreement and four are signatories in process of ratification.[35]

The Moon Agreement is considered as a failure due to the poor number of ratifications and "the fact that major space powers like the USA, China, the Russian Federation, and India have not become Parties has inevitably had an adverse effect on its operational value."[36] Even so, Mexico has signed and ratified the Moon Agreement by its publication on the Mexican Federal Official Gazette on December 27, 1991.[37]

According to the provisions of the Vienna Convention on the Law of Treaties, signing the above international treaties by any State does not constitute a consent to

---

[29]*Supra* note 25.

[30]Mexican Federal Official Gazette. DECRETO por el que se promulga el Convenio sobre la Responsabilidad Internacional por Daños Causados por Objetos Espaciales, aprobado durante el XXVI Periodo Ordinario de Sesiones de la Asamblea General de la Organización de las Naciones Unidas, firmado en las Ciudades de Washington, Londres y Moscú, el 29 de marzo de 1972, August 8, 1974.

[31]*Supra* note 30.

[32]*Supra* note 25.

[33]Mexican Federal Official Gazette. DECRETO por el que se promulga el Convenio sobre el Registro de Objetos Lanzados al Espacio Ultraterrestre, abierto a firma en la ciudad de Nueva York, el 14 de enero de 1975, March 23, 1977.

[34]*Supra* note 30.

[35]*Supra* note 25.

[36]Masson-Zwaan, T., Hofmann, M. (2019). *Introduction to Space Law*. The Netherlands: Wolters Kluwer, p. 36.

[37]Mexican Federal Official Gazette. DECRETO Promulgatorio del acuerdo que debe regir las actividades de los Estados en la Luna y otros cuerpos celestes, December 27, 1991.

be bound by the provisions of a given agreement; signature is just a means to authenticate and express the intent and willingness of the State to conclude the treaty-making process.[38] It is until a State establishes by means of an international act its full consent to be bound by the treaty, that such is considered as ratified.[39] In this sense, Mexico has not only expressed its intent to continue with the treaty-making process by signing international space treaties but has indeed established its full consent to be bound by ratifying all of them. The foregoing can be considered as a demonstration of the intention of the Mexican State to develop and carry out space activities.

4.2 National Law

Mexico is a civil law country that has a formal legal system, where the Mexican Constitution is the supreme set of provisions of the State. Any other legal norm that emanates from it should be in line with the constitutional principles contained therein.

4.2.1 Political Constitution of the United Mexican Sates

The Mexican Constitution does not contain any provision specifically referring to space activities; it is limited to addressing the issue of satellite communication. The Constitution stipulates that satellite communication is to be considered as a priority area within the National Development Plan and that the activities exercised by the State in such sector are not to be considered as monopolies.

Article 28.

The functions exercised exclusively by the State in the strategic areas herein referred shall not constitute monopolies: postal system, telegraphs and radiotelegraphy; radioactive minerals and generation of nuclear energy; the planning and control of the national electric system, as well as the public service of transmission and distribution of electric energy, and the exploration and extraction of oil and the rest of the hydrocarbons, in terms of the sixth and seventh paragraphs of article 27 of this Constitution, respectively; as well as the activities expressly indicated by the laws issued by the Congress of the Union. *The satellite communication* and the railways are *priority areas for the national development* in terms of article 25 of this Constitution; in exercising its governing functions, the State will protect the safety and sovereignty of the Nation, and upon the granting of concessions or permits, it shall preserve or establish the control of the corresponding communication means according to the relevant laws.[40] (emphasis added)

The Mexican Constitution indeed considers satellite communication as a strategic area, but this is only one area within the space sector. Therefore, Article 28 is not a reference for space activities in general. It does not mark a beginning to

---

[38]Vienna Convention on the Law of Treaties, Art. 10, 18.
[39]Ibid., Art. 2,14, and 16.
[40]Political Constitution of the United Mexican States, published on February 5, 1917, Art. 28.

generate legislation corresponding to a broader set of space activities, as the Mexican Constitution only provides for competence regarding satellite communication.[41]

4.2.2 Law that Creates the Mexican Space Agency

The prelude to the creation of the Mexican Space Agency was the now-extinct National Commission of Outer Space (CONEE) created back in 1962, dependent of the Ministry of Communications and Transportation. CONEE was later dissolved in 1977 because the Mexican government did not have medium or long-term objectives that justified its existence.[42]

It was 28 years later, in 2005, that a proposal for the creation of the Mexican Space Agency was presented before the Chamber of Deputies and sent to the Senate for approval. It was not until the year 2008 that the proposal was approved to be transformed into a Law that creates the Mexican Space Agency, hereinafter also referred to as the Law, enacted and published in the Mexican Official Gazette in July 2010.

The Law is an organic law in which the general provisions of the Mexican Space Agency are determined, together with its organization, functioning, budget, and assets. It also adopts a style of corporate governance, with the strengthening of transparency and accountability.[43] Moreover, the Law does not rule for any constitutional principle as there is no such provision contained in the Mexican Constitution related to space activities in general.

The Mexican Space Agency is created as a decentralized public body with legal personality, its own assets and with technical and management autonomy for compliance with its objectives and purposes.[44] However, the Law provides that the Mexican Space Agency will depend on the Ministry of Communications and Transportation for economic purposes, so it is not considered to be fully autonomous.

Article 2 of the Law sets forth the objectives of the Mexican Space Agency, as part of which the Mexican Space Agency has to formulate and propose the General Guidelines of the Mexican Space Policy, as well as the National Program of Space Activities, for approval by the holder of the Ministry of Communications and Transportation. Nonetheless, the carrying out of research and exploration of outer space, as well as the exploitation of celestial bodies or specific space projects, are omitted.[45]

The Mexican Space Agency created working groups in order to discuss and formulate the General Guidelines of the Mexican Space Policy. On July 13, 2011 the General Guidelines of the Mexican Space Policy were published in the Federal

---

[41]López Velarde Sandoval, L.A., (2018). *El espacio exterior y su regulación. Contexto de la actividad mexicana*, Mexico: ECOE Ediciones, p. 132.
[42]Ibid., p. 130.
[43]Ibid., p. 137.
[44]Law that Creates the Mexican Space Agency, published on July 30, 2010, Art.1. Translation provided by the author.
[45]*Supra* note 43.

Official Gazette. These Guidelines consist of thirteen general aspects that the Mexican Space Agency should implement, including the leadership and autonomy of the State in space activities; protection of sovereignty and national security; environmental sustainability; research, scientific and technological innovation and development; development of the productive sector; human resources formation; regulation and certification; international cooperation; financing; disclosure of aerospace activities and their organization and management.[46] The Guidelines are to be reviewed and updated every four years; however, there is no public evidence that they have been reviewed nor updated since their adoption.

The National Program of Space Activities was published in April 2015. It provides a diagnosis about the situation of the space sector in Mexico, concluding that it is less developed in relation to other sectors of the national economy. On the other hand, it establishes objectives, strategies and lines of action to promote the development of the national space sector.[47] The Program aims to set the necessary basis in order to conduct the needed actions for the development of earth observation, global satellite navigation technology, space transportation, satellite communications, applications to improve the efficiency and security of the logistic means, monitoring and surveillance of roads and in general the strategic resources of Mexico.[48]

Surprisingly, in May 2018, the Ministry of Communications and Transportation published the Satellite Policy of the Federal Government,[49] as part of the National Infrastructure Program of the Ministry of Communications and Transportation without the participation and intervention of the Mexican Space Agency and without depending on any of its sectorial programs.

5 Need for Amendments to Mexican Legislation

It is evident that Mexico has not been capable of developing space exploration using its own means. The current situation is clear: for years space activities were not considered a priority for the Mexican government. It was not until the year 2010 that the first legal provisions related to outer space were created and as a result, for more than 30 years, there was no authority in charge of regulating Mexican space activities.

[46]Mexican Federal Official Gazette. ACUERDO mediante el cual se dan a conocer las Líneas Generales de la Política Espacial de México, July 13, 2011.
[47]Mexican Federal Official Gazette. ACUERDO por el que se expide el Programa Nacional de Actividades Espaciales, April 14, 2015.
[48]Ministry of Communications and Transportation. *Programa Nacional de Actividades Espaciales 2013-2018*, February 28, 2019, https://www.gob.mx/sct/documentos/programa-nacional-de-actividades-espaciales-2013-2018 (last accessed on April 11, 2019).
[49]Mexican Federal Official Gazette. ACUERDO que establece la política en materia satelital del Gobierno Federal, May 15, 2018.

The national problem in the space sector was generated, among others, by the lack of a long-term national policy where the guiding factor was the development of space activities and all the benefits derived therefrom.[50]

The current Mexican Space Policy is completely fragmented, since the Mexican Satellite Policy of the Federal Government does not consider the existence of current space policy instruments such as the General Guidelines of the Mexican Space Policy and the National Program of Space Activities, is not supervised by the Mexican Space Agency and treats satellite activity in isolation, although it is by nature contained within the broader notion of space activities.[51]

Economic support and budget allocation to carry out space activities has not been constant and continuous, having a direct effect on Mexican development in outer space. Mexico has been constantly dependent on technologies developed by foreign States in order to launch its satellite systems. Furthermore, there are no specialized human resources that cover the required job profiles of the space sector. The country has a privileged geographical position where launching stations could be built, but this is not being exploited.

Additionally, this lack of national primary legislation triggers another problem: the absence of secondary legislation. The Mexican Judiciary Power has no experience in matters related to the use and exploration of space. Upon the existence of litigation that involves space related issues, the optimal performance of the Mexican judges cannot therefore be guaranteed.[52]

In order to transform space activities into a priority, an integral reform to the existing legislative framework is needed as described below:

(a) Amendment to Article 28 of the Mexican Constitution in order to include space activities as a priority area for national development. This amendment has been previously proposed by other authors such as Mr. Luis Antonio Lopez Velarde Sandoval, which the author of this article completely agrees with. This amendment would constitute the basis for a national set of provisions implementing and regulating space activities. However, this amendment proposal seems to be idealistic from a practical perspective. Mexico has a rigid procedure for constitutional reforms, where the approval of two thirds of the Congress is required as well as approval by the majority of state legislatures. Taking into account that the space sector has not been a priority for the Mexican government and the complex procedure necessary to reform the Mexican Constitution, it is unlikely that this amendment will be achieved.

---

[50]*Supra* note 50.

[51]Ramos Barba, V. *La necesidad de crear una Ley Nacional de Actividades Espaciales para México*, September 4, 2017, https://haciaelespacio.aem.gob.mx/revistadigital/articul.php?interior= 566 (last accessed on April 11, 2019).

[52]Ramirez de Arellano, R.M. (2005). Possible Consequences of the Lack of Secondary Legislation with respect to Outer Space in Mexico in Report of the IISL Space Law Colloquium in Vancouver, Canada, October 2004. *Journal of Space Law, 31(2)*, p. 423–446.

(b) Amendment to the Law that creates the Mexican Space Agency, in order for it to be granted full autonomy and for it to function as an independent body that is not as part of the Ministry of Communications and Transportation.

(c) The creation of a National Law for Space Activities, according to the Political Constitution of the United Mexican States and all international space treaties, containing specific provisions that will allow the operation of space activities in an effective manner, including the faculty to develop and implement the satellite policy as part of Mexican space activities, guaranteeing unified and coordinated management of the use of space objects by governmental entities, in order to optimize the acquisition of satellite applications in an efficient manner from an economic, administrative and social impact point of view.[53]

The abovementioned set of reforms will provide a competitive legal framework which encourages private undertakings to invest within the Mexican State, strengthening commercial development in national space activities. This set of reforms will result "in an increase in jobs and innovations, which is not limited to the space sector."[54]

6 Concluding Remarks

States were the main actors of space activities when the sector first started to develop. However, private entities started to enter into the space market due to several factors. Governments needed financing for space activities and with improvements in technology, the costs of space activities were significantly lowered.

With the development of space activities, legal challenges also arose related to the application of an international legal framework that was initially created for activities which were only executed by governmental entities. States began to create national laws in accordance with international treaties and the specific needs of each country considering the specific space activities they carry out.

Regarding Latin American countries, only Argentina, Brazil and Chile have created national regulations that enable space activities. To be a competitive actor in the space sector, Mexico needs to develop space activities based on a strong legal framework.

Mexico has signed and ratified the five international treaties related to space activities, which can be considered as a first step by the Mexican State to position itself as a leading country in the sector within the Latin American region. However, national legal provisions regarding space activities are not as desired in order to instigate the development of national space activities.

---

[53]*Supra* note 43.
[54]*Supra* note 14, p. 7.

The Mexican Constitution, being the supreme law of the State, does not provide for space activities in general. It limits itself to satellite communication activities as a strategic area. The Law that creates the Mexican Space Agency was recently enacted, providing only for the general rules of existence and functioning of the agency. The Mexican Space Agency is not a fully autonomous body as it directly depends on the Ministry of Communications and Transportation. The Mexican Space Agency has developed the General Guidelines of the Mexican Space Policy and the National Program of Space Activities in order to ensure the development and promotion of Mexican space activities. Mexico has more widely developed satellite infrastructure within the space sector. However, the Mexican Satellite Policy has been published by the Federal Government without intervention from the Mexican Space Agency or any of its sectorial programs.

It is evident that the Mexican Space Policy is divided, the Mexican Satellite Policy has been developed without participation from the Mexican Space Agency, although satellite activities are naturally a part of space activities. On top of that, a long-term policy has not been developed. Governments have failed to continuously support the national space sector, seeing as though there is no formation of specialized human resources and the country has been dependent on foreign technologies for the growth of its space activities.

Space activities need to be considered as a strategic area for the State, in order to create Mexican technology and to avoid reliance on foreign capabilities. With this, the ideal geographic position of the country for space operations could be exploited and thus, Mexico would be able to provide a qualified workforce at competitive prices in an attractive location.

The task for the Mexican government is not easy. However, a good starting point should be the required amendments to the existing Mexican legal framework and the creation of a National Law for Space Activities that will ensure effectiveness in the development and implementation of national space activities, including that of satellite communication. The real challenge is for actors in the Mexican space sector to get the Mexican government to look at space activities and understanding the need for their diversification beyond the existing satellite infrastructure.

The foregoing will result in a national project that guarantees the effective participation of Mexicans in outer space. Said participation will consequently derive in the activation of the market by attracting domestic and foreign investment, which will create jobs and spur innovation in science and technology, together with the complimentary development of the economy in other sectors.

María Clarisa Jiménez Martínez is a professionally qualified lawyer in Mexico pursuing an Advanced LL.M. in Air and Space Law at Leiden University, the Netherlands. The author has six years of professional experience in Corporate Law and Mergers and Acquisitions; as well as extensive experience in finance related operations, specifically structuring and securing aircraft financing transactions.

The Importance of the Aerospace Sector for Mexico: An Industrial, Social and Educative Perspective

Jorge Alfredo Ferrer-Pérez⊙, Carlos Romo-Fuentes, and Rafael Guadalupe Chávez-Moreno⊙

Abstract

Since 2003, aerospace activities had emerged in Mexico starting a new era of opportunities. This work presents how aerospace sector have impacted Mexico from different points of view: industrial, social and educative. First, it is described the aerospace industry, its companies evolution and some key factors that fostered a continuous growth of this sector. It is also acknowledged some of the government strategies to stimulate the aerospace sector. Second, it is mentioned some of the social impacts of the aerospace activities from agriculture to national defense. Third, it is discussed the educative program offer related with aerospace engineering in Mexico. This has a paramount importance since the continuous growth of the sector demands high trained personnel. Finally, conclusions are presented.

1 Introduction

The aerospace sector formed by triple helix players: government, academia and industry, has made important efforts to increase existing capacities and generate conditions that allow the development of this industry at national and regional levels.

J. A. Ferrer-Pérez (✉) · C. Romo-Fuentes · R. G. Chávez-Moreno
Advanced Technology Unit, School of Engineering, UNAM,
Juriquilla Querétaro, Mexico City, Mexico
e-mail: ferrerp@unam.mx

C. Romo-Fuentes
e-mail: carlosrf@unam.mx

R. G. Chávez-Moreno
e-mail: rchavez@comunidad.unam.mx

© Springer Nature Switzerland AG 2020
A. Froehlich (ed.), *Space Fostering Latin American Societies*, Southern Space Studies, https://doi.org/10.1007/978-3-030-38912-3_2

Three specialized corridors have been established in the country (center, northeast and northwest) that place Mexico on the world stage as a viable regional cluster of the aerospace sector, including: infrastructure and existing services, specialized human resources in areas as manufacturing, repair and major maintenance, as well as engineering and design; Moreover, the cost of Mexican labor is considerably lower than in other countries and the proximity to the North American market, made our country an attractive place to invest in the aerospace industry since 2003.

In the last ten years, the Mexican aerospace sector has created a competitive industrial environment in the world and currently Mexico ranks as the third best economy of fourteen analyzed to attract investment in this sector, thanks to its low business costs (labor costs, operation and transportation, a·workforce with high educational level and specific skills, good communications infrastructure and good innovation potential); in general, in Mexico these costs are 21% lower than in the United States.[1]

The growth of this industry has led to the creation of new jobs and companies in the aerospace sector as shown in Fig. 1.[2,3] In recent years, companies increased from 199 in 2009 to 340 in 2017, distributed in seventeen states. In the same period, the generation of jobs grew from 27,000 to 58,000. This represents an annual compound growth rate of 6 and 10%, respectively.

The Ministry of Economy (SE) and the Mexican Federation of the Aerospace Industry (FEMIA), announced the National Strategic Program of the Aerospace Industry "Pro-Aéreo", which remarks the following national objectives:

- To position the country within the first ten places at the international level.
- To foster exports for more than 12,000 million dollars.
- To have 50% of the national integration in manufacturing made by the industry.
- To have a solid index of the industry employment base and encourage its growth.

At the regional level, it was detected that foreign direct investment (FDI) for the aerospace sector from 1999 to the first quarter of 2015 was 3222.4 million dollars.[4]

The main states[5] receiving investment for the sector were: Querétaro with 47.9%, Baja California with 12.9% and Chihuahua with 11.2%. The percentage participation by sector in the manufacture of civil and business aircraft was 27%; manufacture of other components for the aerospace industry, 10%; manufacture of

---

[1]KPMG. (2012). *Competitive Alternatives 2012.* https://assets.kpmg/content/dam/kpmg/pdf/2012/05/Competitive-Alternatives-2012.pdf.

[2]Ornelas, S. L. (2018). MEXICONOW Issue 95. MEXICONOW: https://mexico-now.com/index.php/past-issues/4216-mexiconow-issue-95.

[3]SE. (2012). Secretaría de Economía, Gobierno de México. (FEMIA, & SE, Edits.). Industría y Comercio: http://economia.gob.mx/files/comunidad_negocios/industria_comercio/PROAEREO-12-03-2012.pdf.

[4]Tovar, E. (2015). Modern Machine Shop-México. Industria aeroespacial de México sigue creciendo: https://www.mms-mexico.com/art%C3%ADculos/industria-aeroespacial-de-mxico-sigue-creciendo.

[5]México is formed by 31 states and the capital.

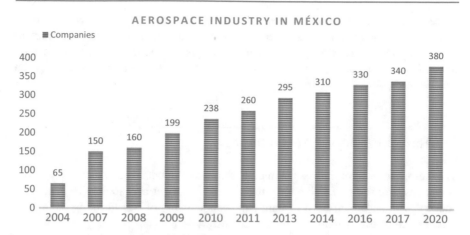

Fig. 1 Evolution in the number of companies in the aerospace sector

cables and electrical components for the aerospace industry, 8% and the remaining 55% corresponded to other items.

These numbers reflect that a good job has been done in the aerospace sector in Mexico, but it is necessary to address at least three essential issues to continue growing and further strengthen the sector:

1. To work closely with other governments and certifications in order to open the opportunities offered by the market, through a competitive industry.
2. To strength the development of the supply chain, in terms of matching the needs of original equipment manufacturers (OEM) and Tier 1 suppliers (TIER 1) with the competencies and capabilities of the productive chains in Mexico and thus generate more attractive aerospace product volumes that lead to better business opportunities.
3. To foster human capital trained in specific skills for companies established in the country. Also, to train engineers capable of establishing new technology-based companies.

Development strategy of the aerospace sector must be based on the generation of cutting-edge technologies, the establishment of new companies, the promotion of clusters and the generation of specialized human capital, where companies come together with universities, research centers and government offices. For this, the strategies designed to achieve national priorities consider harmonizing the current educational plans to the trend of the aerospace sector, especially those plans for undergraduate, master's and doctoral degrees in engineering in areas of control, measurement and advanced materials, offering incentives for exchange and training of human capital between the industry and specialized research centers in the sector, intensifying the training of personnel within the industry in advanced data processing and storage techniques, facilitating the transfer of knowledge from engineering centers and global research, and integrate the talent and infrastructure of laboratories and local research centers with those of the industry to offer spaces for collaboration between the private sector and the government, expanding the execution capacity in the sector.

2 Aerospace Sector Social Impact

The aerospace sector has a paramount relationship with the vision of a more competent and responsible country with its environment and society in general; a country with an urgent need to train engineers capable of transforming laboratory solutions into results with real impact on society.

Mexico, with its megacities and unique distribution of population in its territory, aspires to have engineering solutions capable of formulating pertinent questions to relevant issues, such as the reduction of fuel consumption, noise reduction around airports, replacement of systems of conventional propulsion, mobility in urban areas, reduction of air traffic, electric and hybrid propulsion, additive manufacturing, and many others.

It is difficult to predict the size of the social impact and benefits. For example, research and development of applications around autonomous air mobility, has already been proven and successfully demonstrated in some research groups, using economic energy storage systems that provide greater autonomy, results and social benefit can be very large. The above means autonomous air mobility available to all and transportation to the most inaccessible corners of the national territory. The technology of these autonomous vehicles is complex and implies a great challenge. Among other aspects, an autonomous vehicle is required to be able to communicate in an agile and efficient way, through a navigation and connectivity system throughout the Mexican territory.

Given the challenges posed by new technological developments in order to respond to the demands of society, it is clear the importance of teaching and delving on topics regarding telecommunications and matters related to aerospace engineering. Subjects such as communication systems, automatic control, antennas and radio transceptor systems, avionics, flight mechanics, unmanned aircraft, space communications and others play an important role in solving problems and improving the conditions and optimal development of the various social and educational activities.

The number of systems and components of the aerospace industry that are manufactured and maintained in Mexico is very large. It represents a great opportunity to take advantage of all this platform of infrastructure and jobs that are already in the country, to turn them into services and products of mexican engineers who develop original solutions.

The aerospace industry is a source of employment in many areas and from it a great variety of products and services emerge that affect many vital sectors for the functioning of the modern world, from education, communications, health, the environment and transport, to security and defense. The aerospace industry remains in a continuous movement, innovating and developing new technologies and materials, thus generating displacements in the economic, technological and social sector, thus raising the level of welfare of society.

Nowadays, the applications of aerospace engineering in the world are diverse; In Mexico, for example, the advances that have been made to date have allowed processes such as observation of the national territory, monitoring of climate risks, problems in the field and issues related to national security, among others, to be addressed from a perspective of Aerospace Engineering.

In terms of agricultural production and food self-sufficiency, in developing countries such as Mexico, precision agriculture is a good example of the use of aerospace technology, where instrumentation and observation allow the generation of crop probability maps and crop maps of different types. Products such as corn, agave, avocado, coffee, nuts, wheat, grapes, among others. A direct application is the use of satellites and drones, that allow monitoring various variables in real time to obtain better crops.

Mexican vineyards have benefited from precision farming to produce better wine. With the observation of optical satellites and the use of sensors it is possible to monitor in real time the growth of vines, temperature and irrigation, which allows decisions to be made on the way forward of sowing and subsequent harvesting, without encountering physical in the vineyard. The optical satellites responsible for this task are located 700 km high and take about an hour and a half to travel around the Earth. Unfortunately, Mexico pays large amounts of money abroad for the use of this technology, since it is not its own.

Specifically, Mexico invests around 70 million pesos per year to get access to optical sensors and obtain images from space and proceed to post-processing on Earth.

Only in 2018, around of 3,163 images of satellites were used for Earth observation coming from SPOT 6 and 7, equivalent to 5.8 times the continental area of Mexico for the estimation of the agricultural area of the national territory, data that are of great relevance when applied in the Mexican countryside.

Aerospace technology provides unique opportunities for the creation of useful and profitable goods and services, both public and commercial, with a diversity of activities related to the current problems of our country. There is no doubt that permeating the developments of Aerospace Engineering in sectors such as the Mexican countryside will lead to highly beneficial developments for society, such as converting Mexico from a country that imports food such as corn, to one that exports what it produces reaching a productive independence. The observation by satellites could give us the unbeatable conditions of the harvests and this perhaps, would allow us, also to think about the zero residue.

In relation to disaster prevention, with the help of aerospace technology it is possible to carry out the monitoring of phenomena such as droughts, floods, frosts, pests, hail, among others. The monitoring, image collection, analysis, processing, verification and validation of data, using satellites and observers allows immediate and well-planned actions to be taken, in addition to developing prevention programs for these disasters. In Mexico, thanks to this technology, it is now possible to know that of the 24.6 million hectares of agricultural frontier, 3.6% have a very high risk of drought in the state of San Luis Potosí and that the states of Veracruz and Chiapas are those most affected by the floods. It has been possible to monitor 343,580 hectares of planted area and determine that 207,289 of these were affected by the drought in 2017.[6]

Another example of the use of satellites is in the monitoring of fishing activity. The satellites allow to monitor the closed and fishing areas; For example, in the state of Sinaloa, satellite images and unmanned aerial systems (UAS) from the Pacific Ocean are used, which are received at the ERMEX satellite station (a ground station in Mexico) to track fishing vessels by type. With this technology it is possible to verify that the so-called "pangas" (small boats and with equipment) carry out their activities with the corresponding fishing permits and in the permitted areas, without putting the biodiversity and protected areas of the state at risk.

Regarding the use of aerospace technology in matters of national security, the Secretariat of National Defense (SEDENA) has emphasized that aerospace technology dependence is a risk to national security and constitutes a priority issue for the country.

The role of aerospace technology in security applications is central and can be understood through its three great capabilities: communicate, observe and locate. In this regard, the Mexican government has made important developments in terrestrial communications platforms for national security, with an investment of close to $1,261 million pesos in satellite systems such as MEXSAT, which also considers applications of telemedicine and distance education. This represents a breakthrough in communications[7]; however, it is not enough: the risks of the modern world show the need for better surveillance systems, monitors and prevention to safeguard the security of the population.

Aerospace technology also provides information from territory observation and positioning platforms. This, in addition, represents a great niche for possible Mexican developments in space systems and networks, which respond to national specificities in terms of security. Thus, thanks to the information provided by space platforms for Earth observation, in complementarity with the terrestrial and aeronautical monitoring network, the security of the Mexican territory is benefited. Satellite communications is currently an industry that has been generating wealth

[6]SAGARPA. (2018). Atlas Agroalimentario 2018. Publicaciones SIAP: https://nube.siap.gob.mx/gobmx_publicaciones_siap/pag/2018/Atlas-Agroalimentario-2018.
[7]SCT. (2012). Libro blanco del sistema satelital mexicano para seguridad nacional y cobertura social (MEXSAT): http://www.sct.gob.mx/fileadmin/_migrated/content_uploads/LB_Sistema_Satelital_Mexicano_Mexsat_01.pdf.

for countries that have invested in this sector. National security is one of the applications that has the greatest impact on daily life in the societies of the world.

In the next decade, new implementations will be developed in the aeronautical industry driven by the growing demand for food, security, surveillance, connectivity, among others; therefore, it is vitally important that our country, to face the challenge that comes, be able to train new professionals with a high academic level and sufficient sensitivity to understand the problems that Mexico will face.

Another social need is to promote a culture among young engineers that encourages the development of their own products and services. Having more and more examples of successful cases of products and services developed by Mexican engineers is a phenomena that is permeating the youngest and will transcend the generations.

In all branches of engineering, not only in aerospace, it is necessary to break the barriers that keep Mexico as a manufacturing and service country. The development of products by Mexico has a capital importance to allow the nation to earn its technology independence.

3 Current and Potential Work Field

The market of the aerospace industry in Mexico opens a wide range of work fields where professionals of different engineering branches can be incorporated. Figure 2 shows the commercial sectors that make up this market.

It is important to clarify that the aerospace market also includes ground segment, which is constituted by the telemetry, tracking and command systems of the objects that operate both in the Earth's atmosphere and in outer space.

Similarly, the different commercial activities in the aerospace market can be characterized by manufacture, MRO and R&D and they are presented in Table 1, which also shows their percentage share in economic terms. This is of special interest in the development of the Aerospace Engineering career plan, since it largely identifies the professional activity that will be carried out by the graduate of the race, as well as the skills that should be generated in their comprehensive training.[8]

Derived from the analysis of both the aerospace market and the activities that comprise it, the FEMIA has made a projection of jobs creation in the aerospace industry in Mexico, which is presented in Fig. 3.

It is observed that by 2020 there will be 75,000 jobs in the aerospace industry, of which between 30 and 35% of jobs (38,500 jobs) will occupy by personnel with engineering background in the aerospace field.[9]

---

[8]ProMEXICO. (2017). ProMEXICO: Inversión y Comercio. Mapas de Ruta: http://www.promexico.mx/documentos/mapas-de-ruta/plan-orbita-2.0.pdf.

[9]AI México. (2013). Observatorio de la Ingeniería Mexicana: http://www.observatoriodelaingenieria.org.mx/docs/pdf/5ta.%20Etapa/15.La%20ingenier%C3%ADa%20en%20la%20industria%20aeroespacial%20en%20M%C3%A9xico.pdf.

Fig. 2 Aerospace market

Table 1 Aerospace industry activities in Mexico

Manufacture Aircraft components and parts manufacturing and assembly	MRO Maintenance, Repair y Operations	R&D Research and Development
79%	11%	10%
Harnesses and cables Engine components Landing systems Injection and plastic molds Fuselages Composts Heat exchangers Precision machining	Turbines and engines Fuselages Electrical-electronic systems Landing systems Propellers Dynamic components Coatings, corrosion and protection Interior arrangement and redesign Unitary power systems (APU)	Aerospace dynamics Control systems Flight simulation Non-destructive testing techniques (NDT) Processing of data and images Equipment design Embedded systems

Based on the data reported in 2017 by the National Association of Universities and Institutions of Higher Education (ANUIES), there is a record of 4,523 graduates from different careers with employment in the aerospace sector, of which 82% have

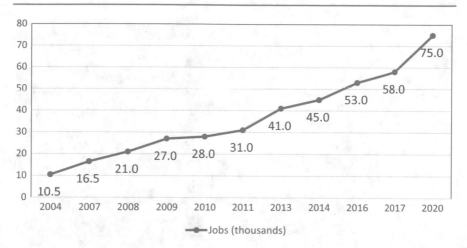

Fig. 3 Jobs in the aerospace sector in Mexico

a bachelor's degree and the remaining 18% have an academic level of associate degree (TSU[10]), a category of studies that is obtained after having completed the high school and represents a higher education modality without reaching the bachelor degree.

In many cases, when graduates are integrated into aerospace sector, companies must additionally train employees through specific courses to have competent personnel, which represents an additional cost to companies.

In this sense, the map of companies and research centers of the aerospace sector in Mexico is presented, which is shown in Fig. 4.[11]

From Fig. 4, the state of Baja California is the entity with the largest number of companies in the aerospace sector in the country, having eighteen states on the list, with a total of three hundred companies and eleven research and development centers as long of the national territory. The turns that these industries have are coatings, design, engineering, research and development, manufacturing, maintenance, repair and revision, testing of materials, non-destructive testing, development of space systems, launch vehicles, modeling of systems and components, emulation of systems and components, among others.

Among the companies that have been identified and established in the country's aerospace market are Aernnova Acrospace México, Axon's Cabling, Eurocopter, Bombardier, Airbus, Indra, General Electric, Honeywell, ITP, Safran, Mexicana MRO, Oaxaca Aerospace, Thales, Boeing, ETU, UTC Aerospace Systems,

[10]TSU is an advance technical degree with a 3 years duration.
[11]Review, M. A. (2017). Mexico Aviation & Aerospace Review 2016/2017. México City: Toguna, S. de R.L. de C.V.

#	ESTADO	COMPAÑIAS	CENTROS I&D
1	Baja California	71	
2	Sonora	52	
3	Querétaro	41	4
4	Chihuahua	35	2
5	Nuevo León	32	2
6	Jalisco	12	1
7	Estado de México	12	
8	Tamaulipas	11	
9	Ciudad de México	10	2
10	Coahuila	7	
11	San Luis Potosí	5	
12	Guanajuato	3	
13	Yucatán	3	
14	Puebla	2	
15	Durango	1	
16	Aguascalientes	1	
17	Zacatecas	1	
18	Hidalgo	1	
	TOTAL	300	11

Fig. 4 Map of companies and research centers of the aerospace sector in Mexico

Edison Effect, Altaser Aerospace, Ketertech, Thumbsat, Composite Solutions, Datiotec Aeroespacial, Advanced Material Solutions, MXSpace y Glenair, among other.

The companies are from micro, medium and small companies (MSMEs) to large transnationals, and there are also companies generated from groups of local entrepreneurs in the aerospace market.

An interesting fact is that 70% of the turnover of the 350,000 million dollars of current aerospace activities derives from the provision of services, while 30% comes from the manufacturing sector. In the services sector we can locate the services provided by satellites, which are a great source of income due to the data derived from them, which are applicable in fields such as transport, agriculture, meteorology, etc..[12]

3.1 Educative Offer

According to the sectoral study conducted by FEMIA and the SE, entitled "Knowing the Aerospace Industry, 2018", which was published by the National Institute of Statistic and Geography (INEGI), it is stated that during the 2016–2017 school year, several higher level degrees were offered which are shown in Table 2.[13]

---

[12]Aviación 21. (5 de junio de 2018). A21. Sector espacial facturó 350 mil mdd en 2017: https://a21.com.mx/index.php/aeroespacial/2018/06/05/sector-espacial-facturo-350-mil-mdd-en-2017.
[13]INEGI. (2018). Portal Único del Gobierno de México. https://www.gob.mx/cms/uploads/attachment/file/315125/conociendo_la_industria_aeroespacial_23mar2018.pdf.

Table 2 Academic programs

State	Degrees offered
Baja California	Bachelor of Science Degree in Aeronautical Engineering
	Bachelor of Science Degree in Aerospace Engineering*
	TSU in Aeronautical Manufacturing (Precision Machining Area)
Chiapas	TSU Pilot
Chihuahua	Bachelor of Science Degree in Aerospace Engineering
	Bachelor of Science Degree in Aeronautical Engineering*
	Bachelor of Business Administration in Airport and Air Business Administration
Ciudad de México	Bachelor of Science Degree in Aeronautical Engineering
Estado de México	Bachelor of Science Degree in Aeronautical Engineering
	TSU in Aeronautical Maintenance (Aircraft Area)
Guanajuato	Bachelor of Science Degree in Aeronautical Engineering
Hidalgo	Bachelor of Science Degree in Aeronautical Engineering
Jalisco	Bachelor of Military Science Pilot Aviator Bachelor of Science Degree in Aerospace Engineering
Nuevo León	Bachelor of Science Degree in Aeronautical Engineering*
Puebla	Bachelor of Science Degree in Aerospace Engineering
Querétaro	Bachelor of Science Degree in Aeronautical Engineering (Manufacturing)
	Bachelor of Science Degree in Aeronautical Engineering (Mechanical Design)
	Bachelor of Science Degree in Electronics Engineering and Control of Aircraft Systems
	TSU in Avionics
	TSU in Aeronautical Maintenance (Aircraft Area)
	TSU in Aeronautical Maintenance (Glider and Motor Area)
	TSU in Aircraft Maintenance
	TSU in Aeronautical Manufacturing (Precision Machining Area)
Sonora	Bachelor of Science Degree in Aeronautical Manufacturing Engineering
	TSU in Aeronautics
	TSU in Aeronautical Manufacturing (Precision Machining Area)
Veracruz	Bachelor of Science Degree in Aeronautical Sciences Engineering

* Denotes academic programs that are offered by different universities at various campuses

It is important to point out that B.S. in Aerospace engineering degree offered in Mexico is strongly focus in aeronautics instead of having a balance between aeronautics and space areas. Likewise, master's degrees in aerospace and aeronautics engineering are offered by different universities across the country.

4 Conclusions

From 2003, the boom of the aerospace industry in Mexico began with the arrival of several prestigious companies which are leaders in the aerospace industry around the world. For 2020 it is expected to have 380 companies generating 75,000 jobs positions. Approximately one third of these positions will be occupy with engineers or professionals with a background related with the aerospace sector. Annually, almost 4,500 new graduates receive a bachelor of associated degree in different aerospace related disciplines. This means that 30,000 jobs positions need to be filled with people coming from other branches of engineering and science. Although Mexico is a country which a notorious reputation as a manufacturer, in the last years there are several efforts related with research and development trying to solve different technological challenges that companies and different universities around the country have. Moreover, new companies related with the space area have been born fostering a more balanced aerospace sector. The key players in Mexico: industry, government and academia are aware of the importance of the aerospace activities for the country, and their common initiatives will keep the growth trend for a win-win outcome.

Dr. Jorge Alfredo Ferrer-Pérez is associate professor of National University Autonomous of Mexico-School of Engineering. He received his Ph.D. in Aerospace and Mechanical Engineering from the University of Notre Dame, South Bend in United States. He is part of the Aerospace Engineering Department and responsible of the Space Propulsion and Thermo-vacuum lab. This facility belongs to the Space and Automotive Engineering National Laboratory located at Juriquilla. His current research areas are nano-heat transfer in solid state devices, thermal control, space propulsion, small satellites and development of space technology.

Dr. Carlos Romo-Fuentes is associate professor of National University Autonomous of Mexico-School of Engineering. He received his Ph.D. in Technical Sciences in the Design of Space Systems considering electromagnetic compatibility criteria from the Aviation Institute of Moscow, Russia. He is part of the Aerospace Engineering Department and responsible of the Electromagnetic Compatibility Laboratory. His current research areas are electromagnetic compatibility, certification tests, space systems and space technology development. Likewise, is the technical responsible of the Space Science and Technology Theme Network from the National Council of Science and Technology from the Government of Mexico.

Dr. Rafael Guadalupe Chávez-Moreno is assistant professor of National University Autonomous of Mexico-School of Engineering. He received his Ph.D. in Mechanical Engineering from the School of Engineering-UNAM. He is part of the Aerospace Engineering Department and responsible of the Model Based on Design lab which belongs to the Space and Automotive Engineering National Laboratory located at Juriquilla. He is an active member of the Mexican Society of Mechanical Engineering and the Space Science and Technology Network. His current research areas include space systems, embedded systems and control systems.

Study and Selection of Satellite Images of Nano Satellites for the Agriculture Field in Bolivia

Puma-Guzman Rosalyn and Soliz Jorge

Abstract

The importance of acquiring terrestrial satellite images for a country is a priority since these images can have different uses such as agriculture, military, cartography, border control, etc. Therefore it is essential for a country like Bolivia to obtain satellite images. According to studies carried out by the Bolivian government, the greatest use of satellite images could occur in agriculture, since it is the fastest growing sector in the country. For this reason, this research is focused on the study of the type of satellite images that would be most useful for the field of agriculture. Because satellite technology was developed in recent years the nanosatellites are increasingly use for different applications. Within these applications, the terrestrial images are the most used nowadays.

This research has two objectives, namely:

(1) The type of images that would be most useful in the field of agriculture in Bolivia.
(2) To select the nanosatellites that can provide these images, taking into account the time of revisit them on the Bolivian territory.

P.-G. Rosalyn (✉)
Industrial Engineering and Systems, Universidad Privada Boliviana, Cochabamba, Bolivia

S. Jorge
Exact Science Department, Universidad Privada Boliviana, Cochabamba, Bolivia
e-mail: jorgesoliz@upb.edu

© Springer Nature Switzerland AG 2020
A. Froehlich (ed.), *Space Fostering Latin American Societies*, Southern Space Studies, https://doi.org/10.1007/978-3-030-38912-3_3

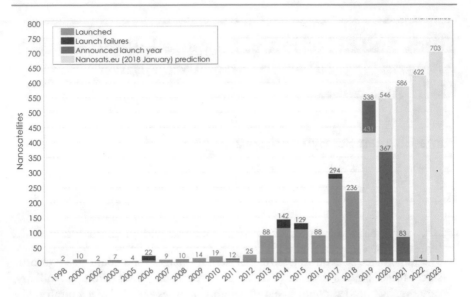

Fig. 1 Nanosatellites launches (Kulu Erik. (2019). Nanosats Database, recovered from https://www.nanosats.eu/#figures)

1 Introduction

Due to the advancements in technology, the possibility of being able to design and build a satellite is more accessible to universities and research groups, that is why, in recent years, the development of pico and nano satellites has been increasing, for this reason space projects are no longer developed only by space agencies (Fig. 1).

The increase in satellite projects is due to the implementation of a standardized platform called cubesat (cube-shaped satellite with dimensions $10 \times 10 \times 10$ cm). In this way, universities and private entities with low budget have the possibility to involve into space missions at a low cost. Within these satellites called cubesats there are configurations of 1 unit ($10 \times 10 \times 10$ cm) 2 units ($10 \times 10 \times 20$ cm), 3 units ($10 \times 10 \times 30$ cm), etc.[1]

Since a cubesat of 1 U cannot carry a camera that generates a good resolution due to limited space, for this reason, configurations of 2 or 3 U that carry a camera for the acquisition of mainly terrestrial images were chosen.

Currently satellites of 2 or 3 U are the most used, as shown in the following graphic (Fig. 2).

To obtain a shorter time of revisit of the satellite over Bolivia, we chose to use a constellation of micro and/or nanosatellites instead of a single larger satellite.

This way we can get more images in a shorter time interval.

---

[1]Hutputtanasin Amy (2004). "Cube Design Specification", revision 9. Cal poly San Luis Obispo.

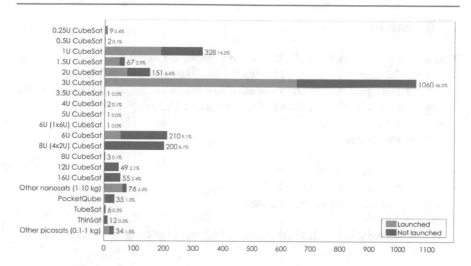

Fig. 2 Nanosatellites types (Kulu Erik. (2019). Nanosats Database, recovered from https://www. nanosats.eu/#figures)

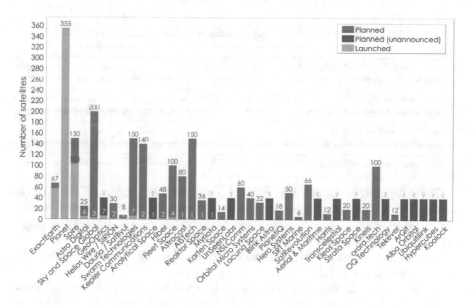

Fig. 3 Nanosatellite constellations (Kulu Erik. (2019). Nanosats Database, recovered from https://www.nanosats.eu/#figures)

The current trend is the launching of constellations of nanosatellites as it can be seen below (Fig. 3).

Based on this trend, our work focuses on identifying constellations of nanosatellites that could provide satellite images of Bolivia to be used in agriculture.

2   Research

This research is focused on identify a constellation of nanosatellites that perform acquisition of terrestrial images. This type of satellites was selected for the low cost of photographs that would be obtained if compared to larger satellites that are currently providing these services to Latin America. This research will only identify these constellations based on:

Evaluation criteria

– Revisit time on Bolivia
– Image resolution as required in the country
– Type of image provided by satellites.

It is worth mentioning that, the missions that meet the evaluation criteria are the ones that will be analyzed.

2.1   Satellites Images Used in Agriculture

Currently the services provided through satellite images for agriculture are:

– Cartography of the vegetal cover of the soil
– Determination of sown areas and inventory of crops by species
– Evaluation of erosive water and wind processes.
– Rapid evaluation of water stress conditions of crops, level of infection or damage of both weeds and insects.
– Delimitation of affected areas, assessment of damages, obtaining general information for the determination of actions; (floods, droughts, tornadoes, hail, etc.).

The images most commonly used in agriculture are those of the satellite LandsatTM, with a spatial resolution of 30 meters, which allows us to detect routes, rivers, urban spots, different types of soil cover and of course differentiate crops, especially in the spectrum of visible and near infrared.

Landsat images with a coverage of 180 km on each side can be purchase, at an approximate cost of $600.

Currently, the satellites and images used in agriculture are 1 m in spatial resolution, in black and white and 4 meters in color resolution, which for the first case means that we can identify any object larger than the square meter, for example: Individual trees, waterways, silos, machines, warehouses, secondary roads, etc.[2]

Until today the experience indicates that the lack of massive consumption of this type of tools is due to poor knowledge of their benefits, as well as their real costs.

---

[2]Ciampagna & Asociados, (2016) Grupo para el Desarrollo de los Sistemas de Información Geográfica. Cordoba, Argentina.

Since the type of satellites and the configuration of the satellites were chosen (constellation), a study of the potential market in Bolivia for the images and of what kind of images would be the most used in the country is required.

2.2 Bolivian Market Study for Satellite Images

The following study of the Bolivian market has been obtained from an official document of the Bolivian State.

2.2.1 Analysis of Demand—Methodology

In order to identify the current use and the potential demand of satellite images in Bolivia, the main sectors of economic activity and government were identified.

These will be the object of this study, that is to say the main sectors within economy, social activity, development, and security in Bolivia. In this sense, the sectors identified are the following:

For each of the defined sectors, both public institutions and private economic actors (in some sectors) were identified. They could be potential users.

Once these institutions were define, the main actors were interviewed for the preparation of this document. Based on the request a document was drawn up to identify current uses of the images and future requirements of both the public and private sectors.

Information was collected about the following topics:

- Current and future use of satellite images
- Identification of applications of satellite systems of Earth Observation, related to the main functions of the institution.
- Identification of projects related to the applications, in addition to the areas of interest for obtaining satellite images required by the institutions.
- Economic situation, infrastructure and human resources of the institutions.
- Identification of the sensors, spatial resolution and periodicity of obtaining images by the institutions based on the applications previously found.

This analysis identifies the need for satellite images for each sector according to the area required by them.

This study and analysis was carried out per sector based on the analysis shown at the beginning of the section (Table 1).

2.2.2 Summary Final Results

The following table presents a summary of the total results of the image areas required by each of the sectors, in this way, the sectors that would make satellite images more useful are identified (Table 2).

From the previous table, it can be concluded that in Bolivia the agricultural sector is the most in need of satellite imagery, where we can also take into account

Table 1 Sectors analyzed to determine the need for satellite images

Sector
Agricultural
Defense and Security
Departmental and Municipal Governments
Hydrocarbons and Energy
Mining and Geology
Planning and Development
Water Resources

Table 2 Area required to be photographed

Sector	Total annual area (km^2)
Agricultural	3.737.258
Defense and Security	1.456.881
Departmental and Municipal Governments	3.298.743
Hydrocarbons and Energy	764.500
Mining and Geology	2.279.162
Planning and Development	1.098.581
Water Resources	6.591.486
Total	19.226.611

that this sector is among the five main items that contribute to the Gross Domestic Product (GDP) of Bolivia, contributing on average, 15% of the national GDP.

Now, this article focuses on the field of greater use of satellite images.

2.2.3 Agricultural Sector

The following list shows the private and public institutions that were consulted for the use of satellite images in Bolivia (Table 3).

Once the information obtained by the previously detailed institutions was processed, we proceeded to verify the requirements of applications required for them (both for public and private institutions).

The following table presents the applications that according to the institutions interviewed would be those required by this sector (Table 4).

As can be seen, the four main applications required by the institutions of the agricultural sector are: Land Use and Coverage, Agriculture, Forestry and Vegetation and Disaster Monitoring, applications that are related to the main functions and activities carried out by the institutions of the sector.

Therefore, in terms of applications, we will focus on agriculture as it was shown to represent the largest field of application in the country.

Table 3 Public and private institutions from Bolivia

Public institution	Private institution
Autoridad de Bosques y Tierras	Asociacion de Productores de Oleaginosas y Trigo
Instituto Nacional de Reforma Agraria	Productores Vitivinicolas Aranjuez Milcast Corp
Servicio Nacional de Sanidad Agropecuaria e Inocuidad Alimentaria	Productores Vitivinicolas y Vinedos "La Cabana" S.R.L
Viceministerio de Desarrollo Rural y Agropecuario	Centro Tecnologico Agropecuario en Bolivia
Viceministerio de Medio Ambiente, Biodiversidad y Cambios Climaticos	Centro Vitivimcola Tarija
Viceministerio de Tierras	Sociedad Agroindustrial del Valle, ltda
Servicio Nacional de Areas Protegidas	Camara Agropecuaria del Oriente
Viceministerio de Coca y Desarrollo Integral	Federacion de Ganaderos de Santa Cruz
	Asociacion Boliviana de Criadores de Cebu
	Asociacion Nacional de Productores de Quinua

Table 4 Applications in the agricultural sector

Application	Number of institutions requiring application	Percentage of institutions requiring application (%)
Use and coverage of floors	16	89
Agriculture	14	78
Forests and vegetation	9	50
Disaster monitoring	9	50
Water resources	7	39
Urban and regional planning	3	17
Environment	2	11
National and regional maps	2	11
National security, defense and surveillance	1	6

Table 5 Surface for specific applications

Sector	Surface (km^2)	Number of photographs required per year
Monitoring of coca crops	6.200	2
Detection of illicit crops	1.098.581	1

Table 6 Analysis of constellations (Kulu Erik (2019). NewSpace Index, recovered from https://www.newspace.im/)

Organization	First launch	Field	Type of image	Revisit times	Sensor resolution	Swath width
Planet	2013	Earth observation	RGB	Variable	3.7 m	Swath of 24.6 km × 16.4 km
Astro digital	2014	Earth observation	RGB and NIR	1 days	22 m	220 km
Dauria/SatByul	2017	Earth observation	Multispectral	16 days	22 m	220 km
Satellogic	2016	Earth observation	Hyperspectral	5 min	1 m	125 km
BlackSky	2016	Earth observation	Panchromatic and multispectral	Few hours or less	1 m	250 km

Table 7 Selected constellations (Kulu Erik (2019). NewSpace Index, recovered from https://www.newspace.im/)

Organization	First launch	Field	Type of image	Revisit times	Sensor resolution	Swath width
Satellogic	2016	Earth observation	Hyperspectral	5 min	1 m	125 km
BlackSky	2016	Earth observation	Panchromatic and multispectral	Few hours or less	1 m	250 km

2.2.4 Application in Agriculture

Focused on agriculture based on the study carried out, three sectors and their surface were detected as requiring periodic monitoring based on (Table 5).

2.2.5 Study of Constellations

Next, the main characteristics of the different constellations of nanosatellites that satisfy the evaluation criteria are shown.

According to this first selection of satellite constellations that meet the indicated criteria, we choose the constellations that have the best characteristics for use in Bolivia.

These evaluations were made based on technical criteria that is to say based on the demand study for Bolivia, and not on the economic aspect.

3 Conclusions

We conclude with this investigation that the use of constellations is required due to the shorter time of revisit over Bolivia, the constellation must be of nano or micro satellites because of the cost perspectives of the images.

As seen in Tables 6 and 7, these constellations are intended for terrestrial observation, the constellations were selected based on their revisit time as well as the type of image that the satellites can provide (agricultural field).

We also conclude that the use of the images is very important for different sectors in the country, if we focus on the agriculture sector it was seen that the use of the images has a great economic potential (as can be seen in the Table 2).

For this reason, in a future work we intend to make a more specific analysis of the images, and its specific application in each part of agriculture for example; the optimization in the use of soils, precision agriculture, etc.

Puma-Guzman Rosalyn is project manager at SUR AEROSPACE (Bolivia) with work experience in USIPSAT project, Cubesat Universidad Simon I Patiño (Bolivia) and CANSAT competition (Bolivia). Her education background: Industrial engineering and systems of the Universidad Privada Boliviana (Bolivia).

Soliz Jorge is professor at the Universidad Privada Boliviana, Exact Science Department (Bolivia) with work experience in space mission analysis, UPCSAT 1 (first satellite of the Universitat Politecnica de Catalunya, Spain), Satellite Libertad 2 (second satellite of Sergio Arboleda University—nano satellite, Colombia). Education background: mechanical engineering of the Universidad Mayor de San Simon (Bolivia) and doctoral program in aerospace science and technology of the Universitat Politecnica de Catalunya (Spain).

CIIIASaT Structure Additive Manufacturing Design

Malena Ley-Bun-Leal, Marlom Arturo Gamboa-Aispuro, Patricia del Carmen Zambrano-Robledo⊙, Ciro Angel Rodriguez-Gonzalez⊙, Omar Eduardo Lopez-Botello⊙, and Barbara Bermúdez-Reyes⊙

Abstract

A three-dimensional Finite Element Model has been developed in order to predict the optimum processing parameters during the laser melting of an AA6061 feedstock using the Additive Manufacturing (AM) process Selective Laser Melting (SLM) for the manufacturing of the CIIIASaT's structure. The presented model computes the heat flow characteristics in three dimensions of the SLM process as well as the melt pool geometry formed during the process. The model considers the phase transformations and physical phenomena present during the SLM of AA6061

M. Ley-Bun-Leal · M. A. Gamboa-Aispuro · P. del Carmen Zambrano-Robledo ·
B. Bermúdez-Reyes (✉)
Facultad de Ingeniería Mecánica y Eléctrica, Universidad Autónoma de Nuevo León,
San Nicolás de los Garza, Nuevo León, Mexico
e-mail: barbara.bermudezry@uanl.edu.mx

M. Ley-Bun-Leal
e-mail: malena.leybun@uanl.edu.mx

M. A. Gamboa-Aispuro
e-mail: marlom.gamboapr@uanl.edu.mx

P. del Carmen Zambrano-Robledo
e-mail: patricia.zambranor@uanl.edu.mx

C. A. Rodriguez-Gonzalez · O. E. Lopez-Botello
Escuela de Ingeniería y Ciencias, Tecnologico de Monterrey, Monterrey,
Nuevo León, Mexico
e-mail: ciro.rodriguez@tec.mx

O. E. Lopez-Botello
e-mail: omarlopez@tec.mx

M. Ley-Bun-Leal · M. A. Gamboa-Aispuro · P. del Carmen Zambrano-Robledo ·
C. A. Rodriguez-Gonzalez · O. E. Lopez-Botello · B. Bermúdez-Reyes
National Laboratory for Additive and Digital Manufacturing (MADiT), Coyoacán, Mexico

© Springer Nature Switzerland AG 2020
A. Froehlich (ed.), *Space Fostering Latin American Societies*, Southern Space
Studies, https://doi.org/10.1007/978-3-030-38912-3_4

in order to predict the thermal gradients present during the process. The developed model was capable of predicting the melt pool size (with an error between 16 and 6% for different samples). Porosity of the manufactured samples was measured as an additional information and for future reference. SLM technologies demonstrated to be capable of producing AA6061 structures for their usage in Cubesats.

1   Introduction

Cubesat satellite structures are an excellent opportunity for emerging countries to dabble into structural design, materials, manufacturing techniques, communication systems, energy, etc. In addition to access a low cost space and train human resources.[1] Cubesats represent a change in the space sector, allowing participation in space activities, as well as promoting scientific and technological development programs.[2] Several universities around the world have designed and manufactured their own cubesat structures in order to generate knowledge in the satellite area.[3]

In Latin America, several universities have developed various missions and placed them intp orbit. The National University of Engineering in Peru developed the cubesat Chasqui in 2012, which its mission was the remote perception of Peru,[4] The Federal University of Santa Maria in conjunction with the National Institute of Space Research of Brazil in 2013, designed the Cubesat NanosatC-BR1 which its mission was the study of disturbances of the magnetosphere.[5] In 2007 and 2016, Sergio Arboleda University in Colombia developed the Libertad 1 and Libertad 2 satellites, in order to monitor the territory.[6] The Technological Institute of Costa Rica designed the Irazú cube, which monitors the country's forest resources.[7]

---

[1]K. Woellert, P. Ehrenfreund, A. J. Ricco, and H. Hertzfeld, "Cubesats: Cost-effective science and technology platforms for emerging and developing nations," Advances in Space Research, vol. 47, pp. 663–684, 2/15/ 2011.
[2]E. Herrera-Arroyave, B. Bermúdez-Reyes, J. A. Ferrer-Pérez, and A. Colín, CubeSat System Structural Design, 67th. International Astronautical Congress (IAC 2016).
[3]M. D. D. Staff, "Small Spacecraft Technology State of the Art," National Aeronautics and Space Administration, California, Technical TP-2014-216648, February 2014.
[4]Renato Miyagusuku, Kleber R. Arias, Elizabeth R. Villota. Hybrid Magnetic Attitude Control System under CubeSat Standards. IEEE 978-1-4577-0557 (2012).
[5]Eduardo Escobar Bürger, Geilson Loureiro, Rubens Zolar Gehlen Bohrer, Lucas Lopes, Cleber Toss Hoffmann, Danny Hernán Zambrano, Guilherme Paul Jaenisch. Development and Analysis of a Brazilian Cubesat Structure. 22nd International Congress of Mechanical Engineering (COBEM 2013).
[6]Garzón A; Villanueva YA (2018) Thermal Analysis of Satellite Libertad 2: a Guide to CubeSat Temperature Prediction. J Aerosp Technol Manag, 10:e4918. https://doi.org/10.5028/jatm.v10.1011.
[7]Marco Gómez Jenkins, Julio Calvo Alvarado, Ana Julieta Calvo, Adolfo Chaves Jiménez, Johan Carvajal Godínez, Alfredo Valverde Salazar, Julio Ramirez Molina, Carlos Alvarado Briceño, Arys Carrasquilla Batista. Irazú: CubeSat Mission Architecture and Development. 67th International Astronautical Congress (IAC2016).

In Mexico, the Autonomous University of Nuevo Leon, began to develop space projects in 2014. Among them is the design of a cubesat-type satellite structure, named CIIIASaT in honor of the research center in which it was designed (Center for Research and Innovation in Engineering Aeronautics). The CIIIASaT is the result of design of a metallic structure for a CubeSat type nanosatellite. Its mission is to test glass-ceramic coatings resistant to cosmic radiation and thermal changes. Although there are commercially available options for this type of structures, they do not completely solve the requirements to meet a specific mission. The CIIIASaT was designed in four phases of design: planning and clarification, conceptual design, preliminary design and detail design. Therefore, for the structural design, structural dynamics were considered due to the loads induced by the launch vehicle, as well as the mechanical stresses to which it will be exposed.

The structural design of the CIIIASaT consists of three components, four integrating shafts and eight Seeger fixed (Fig. 1), which indicates that it has no threaded fastening. Additive Manufacturing (AM) technologies are proposed for the manufacturing of CIIIASaT, specifically using the technology Selective Laser Melting (SLM) and the material AA6061. The latter was selected according to the cubesat standard of CalPoly.[8]

Universities around the world manufacture their cubesat structures in a traditional way. However, over the last 30 years, manufacturing technologies and structural materials applied to space systems have evolved in the form of manufacturing and assembly of structures through AM.[9] Some universities that design cubesat structures according to a specific mission, have used AM to manufacture antenna deployment devices in polymer matrix materials.[10] On the other hand, primary structures of composite material for cubesats have been designed[11,12] and recently SLM has been used for the manufacturing of university cubesats in aluminum alloys.[13,14]

The SLM process is a powder bed technology system, which use an enclosed chamber filled with and inert gas (usually Argon) to prevent oxidation of the different metallic materials. A leveler usually called "wiper" spread a thin layer of

[8]CalPoly-CDS-Rev.13, CubeSat Design Specification, California Polytechnic State University (2013).

[9]Suraj Rawal. Materials and Structures Technology insertion into spacecraft systems: Successes and Challenges. Acta Astronautica 146 (2018) 151–160.

[10]José Antonio Vilan Vilan, Fernando Aguado Agelet, Miguel López Estévez, Alberto Gonzalez Muiño. Flight Results: Reliabity and lifetime of polymeric 3D-printed antenna deployment mechanism installed on Xatcobeo & Humsat-D. Acta Astronautica 107 (2015) 290–300.

[11]Francesca Cuoghi. The Sapec Additive Revolution of CRP USA. Reinforced Plastics. Vol 60, 4, 2016.

[12]P. Gaudenzi, S. Atek, V. Cardini, M. Eugeni, G. Graterol Nisi, L. Lampani, L. Pollice. Revisiting the configuration of small satellite structure in framework of 3D Additive Manufacturing. Acta Astronautica 146 (2018) 249–258.

[13]Scott J. I. Walker, Adrian Dumitrescu. Design for Additive Manufacturing in the Contest of Cubesat Primary Structures. 69th. International Astronautical Congress (IAC 2018).

[14]Alberto Boschetto, Luana Bottini, Marco Eugeni, Valerio Cardini, Gabriel Graterol Nisi, Francesco Veniali, Paolo Gaudezi. Selective Laser Melting of a 1U Cubesat Structure. Design for Additive Manufacturing and assembly. Acta Astronautica 159 (2019) 377–384.

Fig. 1 CIIIASaT structure

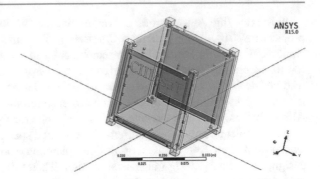

metallic powder with a known thickness over a moving platform. Subsequently, a power source (laser beam) irradiates over the powder melting the material in the shape of the desired geometry. The process repeats layer by layer until the solid 3D part is conformed.[15]

Through numerical models of AM processes it is possible to obtain ideal parameters to optimize the manufacturing of the thin structures required for CubeSats, avoiding waste of materials utilized for tests. These parameters are used to be programmed in the SLM machine and to manufacture the pieces according to the best results obtained in simulations. Later the pieces were analyzed to verify congruence between obtained data and simulated data, and thus demonstrate that AM is highly reliable for obtaining structures used in aerospace industry maintaining quality standards.[16]

Several studies can be found where CubeSat design and manufacturing propose using SLM, but none of them explores the use of AA6061. Thus, the objective of this research is the creation of a numerical model that represents the process using this material in order to predict its thermal history, as well as the model validation through the printing of samples with the proposed parameters. The density of the manufactured samples was measured as additional information.

2   Methodology

2.1   Thermophysical Properties of Solids

Powders processed by AM undergo state changes when they absorb the energy exerted by the laser, changing its state from powder to liquid and from liquid to

---

[15]W.E. Frazier, Metal Additive Manufacturing: A Review, J. Mater. Eng. Perform. 23 (2014) 1917–1928.
[16]T.M. Wischeropp, R. Salazar, D. Herzog, C. Emmelmann, Simulation of the effect of different laser beam intensity profiles on heat distribution in selective laser melting. Lasers in Manufacturing Conference (2015).

Table 1 Composition of AA-6061 (mass %)[a]

Al	Cr	Cu	Fe	Mg	Mn	Si	Ti	Zn
96.45	0.4	0.3	0.7	1	0.15	0.6	0.15	0.25

[a]K.C. Mills, Recommended values of thermophysical properties for selected commercial alloys, Woodhead Publishing (2002)

Table 2 Solid phase thermophysical properties for AA-6061

Temp (K)	ρ (kg/m^3)	C_p (J/K g)	H (J/m^3)	λ (W/K m)
298	2705	0.87	0	167
373	2695	0.95	1.86E + 08	195
473	2675	0.98	4.44E + 08	203
573	2655	1.02	7.06E + 08	211
673	2635	1.06	9.75E + 08	212
773	2610	1.15	1.25E + 09	225
873	2590	1.16	1.54E + 09	200
883	2584.75	1.16	1.56E + 09	196.7
898	2556.75	1.16	1.69E + 09	179.1
910	2471	1.16	2.10E + 09	125.2
915	2415	1.16	2.37E + 09	90
973	2400	1.17	2.52E + 09	91
1073	2372	1.17	2.77E + 09	92

solid. These phase changes are represented by the temperature dependent properties of the material and are implemented in the models in order to simulate the SLM process.

Properties of the solid phase were obtained from experimental data. Mills report the experimental temperature dependent physical properties of the commercial alloy AA6061, commonly used in the main structures of small satellites (CubeSat).[17] The chemical composition of alloy is listed in Table 1.

During a transient thermal analysis, the material thermal properties must be taken into account. Thermophysical properties of liquid phase differ from those of solid phase. Values of density (ρ), thermal conductivity (λ), heat capacity (Cp) and enthalpy (H), become dependent upon the solid fraction at a fixed temperature in the mushy zone. Those values are calculated by Eq. (1), where P_T is the calculated property, $f_{s(T)}$ is solid fraction, $P_{T_{sol}}$ and $P_{T_{liq}}$ the values of the solid and liquid phase.[18]

$$P_T = f_{s(T)} \cdot P_{T_{sol}} + \left(1 - f_{s(T)}\right) \cdot P_{T_{liq}} \tag{1}$$

---

[17]K.C. Mills, Recommended values of thermophysical properties for selected commercial alloys, Woodhead Publishing (2002).

[18]K.C. Mills, Recommended values of thermophysical properties for selected commercial alloys, Woodhead Publishing (2002).

Table 3 Thermophysical properties for AA-6061 powder

T (K)	ρ (kg/m^3)	C_p (J/K g)	H (J/m^3)	λ (W/K m)
298	1487.75	0.87	0	0.2234
373	1516.61	0.95	1.02E + 08	0.9949
473	1582.21	0.98	2.44E + 08	1.84
573	1646.71	1.02	3.88E + 08	2.79
673	1758.44	1.06	5.36E + 08	24.87
773	2240.81	1.15	6.89E + 08	51.74
873	2590	1.16	1.54E + 09	92.24
883	2584.75	1.16	1.56E + 09	196.7
898	2556.75	1.16	1.69E + 09	179.1
910	2471	1.16	2.10E + 09	125.2
915	2415	1.16	2.37E + 09	90
973	2400	1.17	2.52E + 09	91
1073	2372	1.17	2.77E + 09	92

Solid phase thermophysical properties of AA6061 was calculated by Eq. (1) and Mills experimental work, and are listed in Table 2.

2.2 Thermophysical Properties of Powders

Since powder particles are not connected with each other (there are spaces between each powder particle), thermophysical properties differ from the ones of a solid material and must be calculated in order to achieve a model that mimics real life SLM process to a greater extent.

Bulk density of a material in powder form is defined as the relation of dry weight and total volume the material occupy, including the space between each particle and is not and intrinsic property. A measuring cylinder of constant volume and a high precision measurement scale were used in order to measure bulk density at room temperature. The bulk density of AA6061 powder used in this work was 1350.9 kg/m^3, resulting on a Density Ratio of 0.4994 in relation with solid density. Since bulk density is a temperature dependent property, the values at different temperatures until melting temperature were calculated.

Enthalpy is a thermodynamical property defined as the thermal energy flux at constant pressure in a pressure-volume domain. The enthalpy of metallic powders is a result of the relationship between density ratio and solid enthalpy value.

Values of ρ, λ and H for mushy zone and liquid phase for both, powders and solid material, were taken as the same, because as the temperature increases, those properties assume a lineal behavior above the melting point. Thermophysical properties of aluminum powders are shown in Table 3 and were used to define the powder-liquid-solid phase change in numerical model.

2.3 Numerical Model Description

A three-dimensional thermal-transient numerical model was developed to obtain temperatures analysis of the SLM process for AA6061. FEM was created as described by Roberts[19] in ANSYS APDL software. Values of thermophysical properties were defined for substrate, solid material, and powder bed modeled as a porous continuous medium with the properties calculated and the following considerations were taken for model:

- A solid eight-node thermal element was used.
- Convection was not defined.
- A powder bed temperature of 200 °C (473 K) was established.
- An absorption of 0.18 was defined for solid material and 0.05 for liquid phase.
- A layer thickness of 70 µm was set due to the geometry of powders.
- A laser diameter of 70 µm was established since it is the same of machine where the experiment was carried out.

Once the model was conditioned to obtain temperature analysis, several simulations were carried out modifying processing parameters. The first step was define the manufacturing parameters just like shown in Table 4.

2.3.1 Model Volumes

According to the defined bed size and thickness, the program create the necessary points of union (known as keypoints) for the generation of substrate and powder bed. Through keypoints, volumes or solids were created and assigned the corresponding thermophysical properties. Figure 2 shows the volumes created by the keypoints. The image shows how a square was created in the center of the plate, this is because in that area laser will perform thermal energy.

2.3.2 Meshing the Model

The mesh is composed by a number of elements in which the model is divided to calculate temperatures. For the FEM developed in this work, a mesh convergence study was carried out using triangular and quadrangular elements and varying mesh size to determine the maximum time and temperatures reached. Data obtained suggest that the most optimal mesh for model results with 238,411 quadrangular elements.

A picture of the generated mesh for the FEM developed is shown in Fig. 3. It can be seen how elements on the edges of substrate are larger than elements located in center. This is because area of interest for model is the center of substrate, for which a refinement was made in that zone to reduce calculation times.

---

[19]I.A. Roberts, Investigation of residual stresses in the laser melting of metal powders in additive layer manufacturing, Ph.D., University of Wolverhampton (2012).

Table 4 Most outstanding parameters for developed FEM for AA-6061

Simulation	Power (W)	Hatch spacing (μm)	Speed (mm/s)
1	220	100	300
2	220	50	300
3	220	100	150

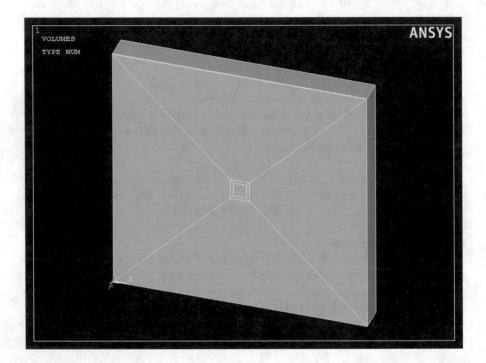

Fig. 2 Volumes created for the FEM implemented in this work

2.3.3 Laser Model

Most of the laser utilized in SLM technologies has a Gaussian profile of intensity, where the center is where the most energy the laser is transmitting to the material. As you move away from the center of the laser spot the energy of the laser beam gets lower and lower.[20,21] The size of center spot and therefore the size of the Gaussian profile, can be changed moving the focus plane of the laser beam (Fig. 4).

[20]G.E. Bean, D.B. Witkin, T.D. McLouth, D.N. Patel, R.J. Zaldivar, Effect of laser focus shift on surface quality and density of Inconel 718 parts produced via selective laser melting, Addit. Manuf. 22 (2018) 207–215.
[21]K.-H. Leitz, P. Singer, A. Plankensteiner, B. Tabernig, H. Kestler, L.S. Sigl, Multi-physical simulation of selective laser melting, Met. Powder Rep. 72 (2017) 331–338.

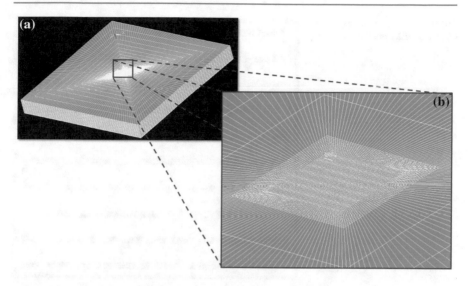

Fig. 3 **a** Isometric view of meshed model, **b** approach to center

Fig. 4 Laser beam Gaussian profile and the relationship with the focus plane (V. Algara Muñoz, Analysis of the optimal parameters for 3D printing aluminum parts with a SLM 280 machine, Universitat Politècnica de Catalunya, 2017.)

Laser was modeled as a mobile heat energy source with Gaussian temperature distribution. The model is described as a continuous heat flow as shown in Eq. (2), where q_L is heat flux of laser, P_L power of laser, A the absorption coefficient of material, r radius of laser, V_w laser speed, and t elapsed time.[22]

$$q_L = \frac{P_L \cdot A}{\pi \cdot r^2} \cdot e^{\left(-\frac{(x + V_w \cdot t)^2 + y^2}{r^2}\right)} \tag{2}$$

---

[22]I. Tomashchuk, I. Bendaoud, P. Sallamand, E. Cicala, S. Lafaye, M. Almuneau, Multiphysical modelling of keyhole formation during dissimilar laser welding, (2016) 7.

Fig. 5 Meander strategy
used for FEM model

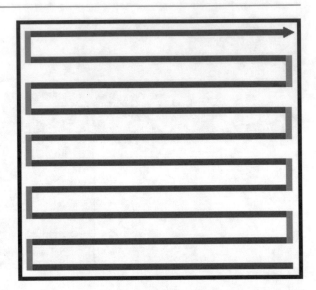

Laser was configured to follow a "meander" scan path, as shown in Fig. 5. Strategy consists of starting at one of the corners of the piece and traveling in the form of a coil in the area of interest until covering the entire volume of the piece.

2.4 Experimental Methodology

In order to predict optimal processing parameters for AA6061 a DOE was produce. The DOE was undertaken with the assistance of Minitab 18, using a 2 level full factorial design with 3 factors Laser Power (P), Laser Scanning Speed (S) and Hatch Spacing (HS) and one response (Relative Density). A full factorial design was selected because its ability to assess the interaction among all the factors. The low and high levels were chosen based on parameters reported on the literature.[23] The value of the layer thickness was fixed at 70 μm according to the powder size distribution. The values of P and S were adjusted in order to avoid the cracking previously observed.

The classic 2^3 factorial design suggested nine experiments including a center point (Table 5). Two specimens were fabricated for each experiment with a total of 18 samples. The specimens were produce on a SLM Solutions 280HL system. Each sample was 20 × 20 × 3 mm and build in the chamber as shown in Fig. 6.

[23] A.H. Maamoun, Y.F. Xue, M.A. Elbestawi, S.C. Veldhuis, Effect of SLM Process Parameters on the Quality of Al Alloy Parts; Part I: Powder Characterization, Density, Surface Roughness, and Dimensional Accuracy, Preprints, 2018.

Table 5 Process parameters used for building the AA-6061 samples

Sample	P (W)	S (mm/s)	HS (mm)	Relative density
1	180	150	0.05	94.15
2	220	150	0.05	93.57
3	180	300	0.05	93.30
4	220	300	0.05	94.24
5	180	150	0.1	94.30
6	220	150	0.1	94.51
7	180	300	0.1	93.06
8	220	300	0.1	94.63
9	200	225	0.075	94.55

Fig. 6 Produced samples of AA-6061 in SLM 280 HL machine

The powder used was a custom made AA6061 alloy, with a particle size of 23–70 μm. The morphology of the powder detected using Scaning Electron Microscopy (SEM) is shown in Fig. 7. Particle diameter was measure in situ to validate the particle size given by the fabricator.

Density measurements were made on as-built samples via water displacement by Archimedes method, according to ASTM B962-17. Six of the specimens were used to developing the numerical model and three were use as validation samples.

In order to validate the model, metallographies of the specimens were obtained. Each specimen were mounted on bakelite, grinded and polished until mirror finish with 1 μm diamond paste. Once the samples were polished, they were etched with Weck's reagent (100 mL H_2O, 4 g $KMnO_4$, 1 g NaOH) and examined in order to measure the melt pool size of corresponding specimens using the image processing software ImageJ. Each sample was immersed for 90 s and 20 measures of the melt

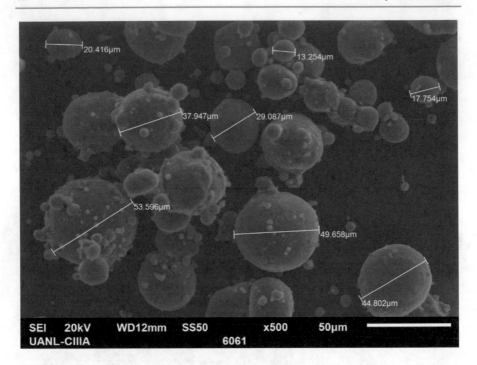

Fig. 7 SEM observations of the powder morphology and size

pool size were taken. The metallography analysis was made with a Zeiss Axio Observer Z1m inverted microscope using polarized light.

3 Results

The powder morphology observed is mostly spherical, but presents a high number of satellites. The satellites can cause difficulties in the processing of the material because they may reduce the powder flow ability and result in a low density powder bed, thus affect the density of the final piece.[24] All the measured particles were in the range between 23–70 μm.

When performing the simulations with variation of parameters, different temperature gradients were obtained in the model resulting in some models with unfavorable results, so only the most significant results for this material are presented. The maximum temperatures reached with the established parameters are presented in Table 6. It can be seen that the temperature ranges are below the

---

[24]Jun Hao Tan, W.L.E. Wong, K.W. Dalgarno, An overview of powder granulometry on feedstock and part performance in the selective laser melting process, Addit. Manuf. 18 (2017) 228–255.

Table 6 Maximum temperatures reached in FEM models

Sample	P (W)	HS (mm)	S (mm/s)	Max. temp. (K)
4	220	100	300	2668.9
6	220	50	300	2712.1
8	220	100	150	2690.3

Table 7 Average diameters for numerical models and experimental samples

Sample	Numerical model		Experimental data	
	Average diameter (μm)	Average depth (μm)	Average diameter (μm)	Average depth (μm)
4	223	112	260	95
6	374	144	408	140
8	334	174	384	165

boiling point of AA6061 (approximately 2793 K) which means the material can be subjected to the manufacturing process without complications.

The melt pools of numerical models and experimental samples were measured by ImageJ and the averages of melt pool sizes are shown in Table 7.

A difference between measured and simulated can be observed, but an error of 16% is obtained for sample 4, a 6% for sample 6, and 9% for sample 8, obtaining similar results to Foroozmehr et al.[25] This means that it is necessary to optimize the models for a more suited prediction of the behavior of the material in manufacturing process.

In Fig. 8 one of the microscopies made to sample 4 can be observed. The blue tonality of the sample is due to the fact that it was etched with Weck's reagent and polarized light was used to observe melt pools more easily. Some of the pools detected and used for the measurements are highlighted.

In the same way, an image taken from numerical model is presented to make the predictions of the fusion pools in Fig. 9. The critical temperatures of the process were set, such as temperature of the powder bed (blue), mushy zone temperature (green) and liquid temperature of aluminum (red) discussed above.

A microscope image of sample 6 can be seen in Fig. 10. Like the previous one, a blue hue is present due to the reagent used to reveal the melt pools. It is possible to notice a pool size greater that sample 1, mainly due to the reduction of Hatch Spacing resulting in a higher heat transfer by the reduction of the time in which the laser travels the nearby area.

Figure 11 shows a capture of melt pool obtained in a time interval of the simulation. Similar to the experimental results, a wider and deeper pool than sample 4 is observed.

---

[25]A. Foroozmehr, M. Badrossamay, E. Foroozmehr, S. Golabi, Finite Element Simulation of Selective Laser Melting process considering Optical Penetration Depth of laser in powder bed, Mater. Des. 89 (2016) 255–263.

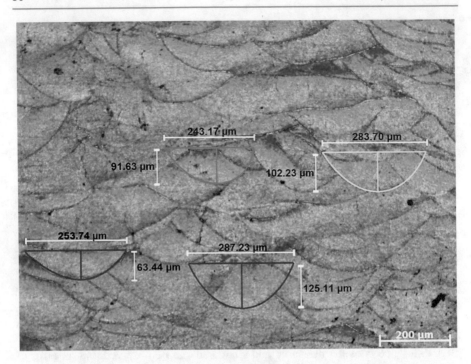

Fig. 8 Microscopy of sample 4 etched with Weck's reagent and melt pool measurements

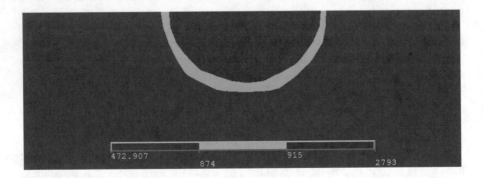

Fig. 9 Melt pool obtained from numerical model of sample 4

Finally, an image obtained from microscope for sample 8 is shown in Fig. 12. There is an increase in depth of the pools compared to samples 4 and 6. This is mainly due to the reduction in the scan speed configured for the manufacture of this sample.

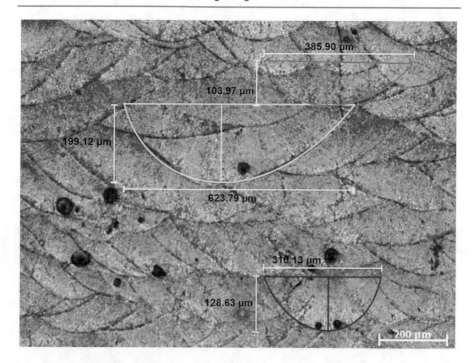

Fig. 10 Microscopy of sample 6 etched with Weck's reagent and melt pool measurements

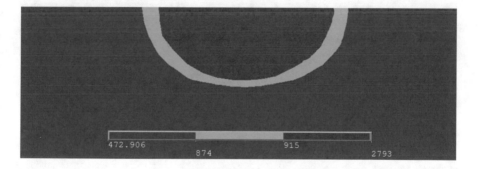

Fig. 11 Melt pool obtained from numerical model of sample 6

Figure 13 shows one of the pools used for the validation of numerical model developed for sample 8. Like the one obtained from the sample manufactured, a pool with a shape similar to half ellipse is shown due to the energy of laser is greater when reducing scan speed.

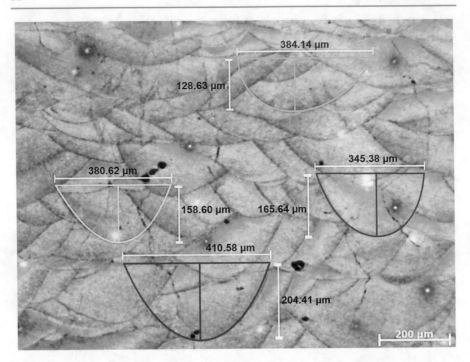

Fig. 12 Microscopy of sample 8 etched with Weck's reagent and melt pool measurements

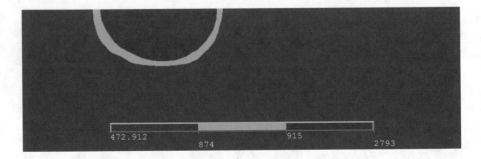

Fig. 13 Melt pool obtained from numerical model of sample 8

4   Conclusions

Several numerical models with a variation of manufacturing parameters were made in order to obtain a temperature analysis of the SLM process for AA6061. Through the simulations developed, it was possible to predict the size of melt pools of manufactured samples, obtaining a maximum error of 16% between experimental samples and numerical models.

It is important to emphasize that no critical temperatures were reached for the material, so no overheating was observed in areas where laser beam had previously been affected. The data obtained from numerical models were used for comparison with samples manufactured in AA6061 and thus validate the predictions of simulations.

Experimental samples data will be used in future research in order to develop a more efficient numerical model. Further density analysis and simulation data will be carry on for process parameters optimization of CIIIASaT material structure, given that the maximal relative density value as-build reached in this work was 94.63%. Skin parameters for AA6061 may be the cause of this density value, so optimization of surface is considered for future work.

Malena Ley-Bun-Leal Master student in Aeronautical Engineering with orientation to structures of Centro de Investigación e Innovación en Ingeniería Aeronáutica of Facultad de Ingeniería Mecánica y Eléctrica from Universidad Autónoma de Nuevo León. Electromechanical Engineer with specialty in automation graduated from Instituto Tecnológico de Los Mochis. Her research lines are: Additive Manufacturing, Aerospace Structures.

Marlom Arturo Gamboa-Aispuro is a M.S. student in Aeronautical Engineering with orientation to structures of Centro de Investigación e Innovación en Ingeniería Aeronáutica of Facultad de Ingeniería Mecánica y Eléctrica from Universidad Autónoma de Nuevo León. Electromechanical Engineer with specialty in automation graduated from Instituto Tecnológico de Los Mochis. He currently works in research lines such as Additive Manufacturing and Aerospace Structures.

Patricia del Carmen Zambrano-Robledo is Senior Research Professor in the Research and Innovation Center in Aeronautical Engineering of Autonomous University of Nuevo Leon. She is faculty member of the Materials Science Doctorate program and Aeronautical Doctorate Program. Her research areas of interest include the processing and characterization of metal- and ceramic-matrix composites (MMCs and CMCs), as well as of superalloys and Additive Manufacturing. She has published around 50 papers in international journals and about 100 articles in international conferences. She is the author or coauthor of about 6 book chapters. Dr. Zambrano Robledo is the organizer of Woman in Science in Monterrey NL, and Infant Spring Course. She has led more than 20 linked projects with industries like AIRBUS and MD Helicopters. Since January 2016 she is a Research Director in UANL. She is currently vice president of the Mexican society of materials (2019–2020) and is the president-elect for the period 2021–2022.

Prof. Ciro Angel Rodriguez-Gonzalez holds a B.S. in Mechanical Engineering from The University of Texas at Austin (1989) and a Ph.D. from The Ohio State University (1997). He worked in the automotive and machine tool industry. He was Staff Engineer at the Engineering Research Center for Net Shape Manufacturing in the USA. He is Visiting Scientist at the Osteo engineering Lab of The Ohio State University (2016–2018). He is currently Professor at the Department of Mechanical Engineering and Advances Materials at Tecnológico de Monterrey. He leads the research group in advanced manufacturing. His research interests include additive manufacturing, micro manufacturing, tissue engineering and technology innovation. He has been Principal Investigator of various research projects with funding from government agencies and industry, with total awards exceeding $2.5 million USD. He is member of the National System of Researchers (CONACyT) with Level 2. Prof. Rodríguez has given invited lectures and courses in Colombia, Costa Rica, El Salvador, USA, Italy, Venezuela and Mexico. He has been visiting scientist and teacher in Brazil, Costa Rica, Cuba, Spain, USA, Italy, and Portugal.

Omar Eduardo Lopez-Botello PhD in Mechanical Engineering specialised in Advanced Additive Manufacturing at the University of Sheffield in the UK, MSc in Materials Science and bachelor degree in Mechatronics Engineering, both from the Autonomous University of Nuevo León. Researcher at Tecnológico de Monterrey, and previously at the Innovation and Research Centre of Aeronautical Engineering of the Mechanical and Electrical Engineering Faculty of the Autonomous University of Nuevo Leon. He forms part of the Advanced Manufacturing research group of Tecnológico de Monterrey. During 2016 he was research associate in the Advance Additive Manufacturing Centre at the University of Sheffield, where he developed a new metal additive manufacturing process. He constantly participates in research projects with industry. His research areas include, development of new additive manufacturing processes, development of new materials for additive manufacturing and numerical modelling of manufacturing processes. He has been awarded with candidate level by the National Researcher System of CONACYT.

Barbara Bermúdez-Reyes Research Professor at the Center for Research and Innovation in Aeronautical Engineering of the Faculty of Mechanical and Electrical Engineering of the Autonomous University of Nuevo Leon. PhD in Sciences of Metallurgy and Materials Science from the Michoacana University of San Nicolas de Hidalgo. She is General Secretary of the University Space Engineering Consortium-Mexico Chapter for the period 2018–2010. She is member of Space Science and Technology Network. She has collaborated in projects with Mexican Air Force and has evaluated projects of Mexican Naval Research Institute. She is part of organizing committee of 5th National Cansat Contest 2019. Her research lines are: Design and processing of aerospace ceramic-ceramic and metal-ceramic composites materials and protective coatings for satellite structures.

Using Space Technology to Mitigate the Risk Caused by Chronic Kidney Disease of Unknown Etiology (CKDu)

Jörg Rapp, Engelbert Niehaus, Alexandre Ribó, Roberto Mejía, Edgar Quinteros, and Anna Fath

Abstract

Chronic Kidney Disease of unknown etiology (CKDu) is a severe problem in rural areas in developing countries. In El Salvador it is the fifth leading cause of death among the adult population. The etiology of the disease is unknown, suspected risk factors are inter alia the exposition to pesticides and to nephrotoxic heavy metals. In this publication we demonstrate an approach, how space technology can be used to mitigate the risk for CKDu. In this approach space technology is used in two ways: to determine the geolocation of a risk exposed person and to gain remote sensed data. Together with a Spatial Decision Support System (SDSS), the geolocation of a risk exposed person is used to deliver risk mitigation strategies tailored to the demand of the user. Remote sensed data (multi-spectral images) are used to create NDVI maps with the aim, to adjust agrochemical application rates to the only needed demand.

J. Rapp (✉) · E. Niehaus · A. Fath
Institute for Mathematics, University Koblenz-Landau, Landau, Germany
e-mail: rapp@uni-landau.de

E. Niehaus
e-mail: niehaus@uni-landau.de

A. Fath
e-mail: fath@uni-landau.de

A. Ribó
SOTASÒL—Serveis de Geologia, Barcelona, Catalunya, Spain

R. Mejía · E. Quinteros
Ministry of Health El Salvador, National Institute of Health, Urbanización Lomas de Altamira, San Salvador, El Salvador

© Springer Nature Switzerland AG 2020
A. Froehlich (ed.), *Space Fostering Latin American Societies*, Southern Space Studies, https://doi.org/10.1007/978-3-030-38912-3_5

1 CKDu in El Salvador and Related Research

1.1 Introduction: Chronic Kidney Disease of Unknown Etiology (CKDu)

In the last decades an increasing number of people suffering on Chronic Kidney Disease (CKD) can be observed in rural areas in developing countries.[1] CKD is a bundle of symptoms caused by different diseases and is diagnosed if "abnormalities of kidney function or structure is present for more than 3 months".[2] Traditional risk factors for the occurrence of CKD in the developed world are inter alia an unhealthy live style, diabetes mellitus, hypertension[3] and obesity.[4] However, these traditional risk factors can be excluded in rural areas in developing countries, where a high prevalence of CKD was observed in the last decades.[5] Due to the unknown etiology, CKD is called Chronic Kidney Disease of unknown etiology (CKDu) in these regions. In recent research CKDu is also named as chronic interstitial nephritis in agricultural communities (CINAC),[6] Mesoamerican Nephropathy (MeN)[7] or Nephropathy of Unknown Cause in Agricultural Laborers (NUCAL).[8] Countries affected by CKDu are inter alia El Salvador, Nicaragua, Belize, Mexico, Sri Lanka, India, Tunisia and Egypt.[9] According to Jayasinghe, the most people affected by CKDu are living in poor, rural communities.[10] In 2013, the existence of CKD

[1]Lunyera, J., Mohottige, D., Von Isenburg, M., Jeuland, M., Patel, U. D., & Stanifer, J. W. (2016). CKD of uncertain etiology: a systematic review. *Clinical Journal of the American Society of Nephrology*, *11*(3), 379–385.

[2]The Renal Association (2019, June, 02). *Stages of CKD*. Retrieved from https://renal.org/information-resources/the-uk-eckd-guide/ckd-stages/.

[3]Barsoum, R. S. (2006). Chronic kidney disease in the developing world. *New England Journal of Medicine*, *354*(10), 997–999.

[4]Levey, A. S., Coresh, J., Bolton, K., Culleton, B., Harvey, K. S., Ikizler, T. A., ... & Levin, A. (2002). K/DOQI clinical practice guidelines for chronic kidney disease: evaluation, classification, and stratification. *American Journal of Kidney Diseases*, *39*(2 SUPPL. 1).

[5]Orantes, C. M., Herrera, R., Almaguer, M., Brizuela, E.G., Hernández, C. E., Bayarre, H., ... & Velázquez, M. E. (2011). Chronic kidney disease and associated risk factors in the Bajo Lempa region of El Salvador: Nefrolempa study, 2009. *MEDICC review*, *13*, 14–22.

[6]Orantes-Navarro, C. M., Herrera-Valdés, R., Almaguer-López, M., Lopez-Marin, L., Vela-Parada, X. F., Hernandez-Cuchillas, M., & Barba, L. M. (2017). Toward a comprehensive hypothesis of chronic interstitial nephritis in agricultural communities. *Advances in chronic kidney disease*, *24*(2), 101–106.

[7]Wijkström, J., Leiva, R., Elinder, C. G., Leiva, S., Trujillo, Z., Trujillo, L., ... & Wernerson, A. (2013). Clinical and pathological characterization of Mesoamerican nephropathy: a new kidney disease in Central America. *American Journal of Kidney Diseases*, *62*(5), 908–918.

[8]Subramanian, S., & Javaid, M. M. (2017). Kidney disease of unknown cause in agricultural laborers (KDUCAL) is a better term to describe regional and endemic kidney diseases such as Uddanam nephropathy. *American Journal of Kidney Diseases*, *69*(4), 552.

[9]Hoy, W., & Ordunez, P. (2017). Epidemic of Chronic Kidney Disease in agricultural communities in Central America. *Pan-American Health Organisation (PAHO)*.

[10]Jayasinghe, S. (2014). Chronic kidney disease of unknown etiology should be renamed chronic agrochemical nephropathy. *MEDICC review*, *16*(2), 72–74.

affected agricultural communities in Central America was recognized at the 152nd Executive Committee of the Pan American Health Organisation (PAHO).[11]

In the scientific community there are different hypothesis available about factors being responsible for the occurrence of CKDu. According to Soderland, Lovekar, Weiner, Brooks and Kaufman, suspected risk factors are high temperature, hard working conditions and related dehydration as well as exposition to nephrotoxic substances, heavy metals and pesticides. However, a concrete evidence for one of the mentioned factors could not be found. Nowadays, also multi-factorial models are used to describe the epidemic of CKDu.[12] For example Jayasumana, Gunatilake and Senanayake proposed a multi-factorial model, in which hard water, the exposition to nephrotoxic metals and the exposition to the pesticide glyphosate are responsible for CKDu, whereby a person is affected by CKDu, if he or she is exposed to all of the mentioned factors.[13]

El Salvador is one of the countries with the highest prevalence of CKDu and it is nowadays a severe problem and the fifth leading causes of death among the adult population.[14] In El Salvador, CKDu was first described almost 20 years ago.[15] However, it took around 10 additional years, until research about that disease was carried out in Central America.[16] At the same time research related to CKDu started also in Sri Lanka.[17] In El Salvador CKDu is characterized by tubulointerstitial nephropathy and moreover by vascular, genitourinary and neurological abnormalities. People typically affected by CKDu in Central America and Asia are young to middle aged male farmworkers, but also women and children are affected. To describe the etiology, the authors proposed a multifactorial model including pesticide exposure, heat stress, dehydration, environmental contaminants, effects of

---

[11]Pan American Health Organisation (PAHO) (2019, June, 05): *La Enfermedad renal crónica de las comunidades agrícolas de Centroamérica es reconocida por la OPS*
Retrieved from https://www.paho.org/els/index.php?option=com_content&view=article&id= 819:la-enfermedad-renal-cronica-comunidades-agricolas-centroamerica-reconocida-ops&Itemid = 291.

[12]Soderland, P., Lovekar, S., Weiner, D. E., Brooks, D. R., & Kaufman, J. S. (2010). Chronic kidney disease associated with environmental toxins and exposures. *Advances in chronic kidney disease, 17*(3), 254–264.

[13]Jayasumana, C., Gunatilake, S., & Senanayake, P. (2014). Glyphosate, hard water and nephrotoxic metals: are they the culprits behind the epidemic of chronic kidney disease of unknown etiology in Sri Lanka?. *International journal of environmental research and public health, 11*(2), 2125–2147.

[14]See Footnote 5.

[15]Trabanino, R. G., Aguilar, R., Silva, C. R., Mercado, M. O., & Merino, R. L. (2002). End-stage renal disease among patients in a referral hospital in El Salvador. *Revista panamericana de salud publica = Pan American journal of public health, 12*(3), 202–206.

[16]Wesseling, C., Crowe, J., Hogstedt, C., Jakobsson, K., Lucas, R., & Wegman, D. H. (2013). The epidemic of chronic kidney disease of unknown etiology in Mesoamerica: a call for interdisciplinary research and action.

[17]Jayasumana, C., Paranagama, P. A., Amarasinghe, M. D., Wijewardane, K. M. R. C., Dahanayake, K. S., Fonseka, S. I., … & Senanayake, V. K. (2013). Possible link of chronic arsenic toxicity with chronic kidney disease of unknown etiology in Sri Lanka.

low socioeconomic status, and genetic susceptibility.[18] It is very plausible that environmental exposure to toxins has a preponderant role in the genesis of CKDu: Orantes et al. showed that renal damage begins frequently in childhood and is often caused by environmental toxins.[19]

In the following section we will give an overview about environmental research related to CKDu carried out in El Salvador.

1.2 Environmental Research

In El Salvador, different studies about the presence of factors probably being responsible for CKDu were carried out. The most studies have the focus on agrochemical use and exposition to heavy metals. The focus on research about pesticide use and heavy metals has the origin in the traditional high pesticide application rates and related stockpiles of pesticides in El Salvador as well as in the geological characteristics in El Salvador. VanDervort, López, Orantes and Rodríguez analyzed the influence of different parameters to the occurrence of CKDu in El Salvador. The study indicates, that high CKDu rates can be especially observed in regions with high agricultural activity and in regions with high pesticide use. Ambient temperature does not have such a high influence like agricultural activity.[20]

Mejia et al. analyzed the legislation for and the hazard of pesticides used in El Salvador. In El Salvador, there is an incomplete pesticide legislation and a poor law enforcement to prevent the misuse of pesticides. Some hazardous pesticides legal in El Salvador are banned or restricted in the European Union and are recommended to be banned by the Rotterdam Convention.[21]

In another study the presence and exposure to pesticides in the rural community Loma de Gallo in El Salvador was analyzed. Loma de Gallo is a farming community located in the lowlands of El Salvador in a sugarcane production area and close to a former pesticide factory with a CKDu prevalence near to 25%. Its inhabitants have different exposition patterns to pesticides: local farmers mishandle pesticides and do not understand the right usage, massive application of pesticides to sugarcane fields by aircraft, drinking water with high levels of pollution from natural and anthropic sources. But also stockpiles of hazardous obsolete pesticides play an important role: presence of stockpiles during the last 30 years in poor

[18]See Footnote 6.

[19]Orantes-Navarro, C. M., Herrera-Valdés, R., Almaguer-López, M., Brizuela-Díaz, E.G., Alvarado-Ascencio, N. P., Fuentes-de Morales, E. J., ... & Zelaya-Quezada, S. M. (2016). Chronic kidney disease in children and adolescents in Salvadoran Farming Communities: NefroSalva Pediatric Study (2009-2011). *MEDICC Review*, *18*, 15–21.

[20]VanDervort, D. R., López, D. L., Orantes Navarro, C. M., & Rodríguez, D. S. (2014). Spatial distribution of unspecified chronic kidney disease in El Salvador by crop area cultivated and ambient temperature. MEDICC review, 16(2), 31–38.

[21]Mejía, R., Quinteros, E., López, A., Ribó, A., Cedillos, H., Orantes, C. M., ... & López, D. L. (2014). Pesticide-handling practices in agriculture in El Salvador: an example from 42 patient farmers with chronic kidney disease in the Bajo Lempa region. Occupational Diseases and Environmental Medicine, 2(03), 56.

conditions (without any monitoring) in a former pesticide factory located close to the community with scarce access restrictions. After the withdrawal of the stock-piles the pollution still remains in the former factory facilities and surrounding areas.[22]

Results from a study conducted by Azaroff and Neas support the hypothesis that members of farmer families, including children, can be exposed to levels of pes-ticides sufficient to be associated with acute health effects even if they do not perform fieldwork.[23] In rural areas, the contamination of human milk by pesticides was identified in the early eighties in El Salvador. It is a persistent problem due to the abuse and mishandling of pesticides.[24]

Furthermore a former pesticide factory is located at the east side of El Salvador, where a community is additionally exposed to drinking water with high levels of pollution from natural (volcanic) and human sources (pesticides).[25]

The high exposition to agrochemicals in El Salvador can be explained by the agricultural history of the country. During major parts of the 20th century, El Salvador, as the other Central American countries, was a cotton producer. This was a mono-cultural cultivation that occupied most of lowlands and coastal areas. As other producer countries, pesticides were massively used in the cultivation because cotton has a high vulnerability to pests. First, at the beginning of the 20th century, fine dust of calcium arsenate was applied. During the fifties the insecticide DDT was widely used against cotton pests and as vector control of mosquitoes for malaria risk mitigation. Together with DDT other organochlorine pesticides, like toxaphene, were used. Pesticide use rose because pests developed resistances and new pesticides like organophosphates were applied together with higher doses of old ones. This massive application was followed by high rates of acute pesticide poisoning and pesticide related illness. However, most of the long term effects were not well recognized because of the low availability of health service delivery in rural areas.[26]

Finally, the cotton crisis of the late seventies and the beginning of the civil war in El Salvador produced a radical reduction of cotton cultivation areas. Production facilities and pesticide factories were abandoned without any supervision.[27] After

---

[22]Quinteros, E., Ribó, A., Mejía, R., López, A., Belteton, W., Comandari, A., … & López, D. L. (2016). Heavy metals and pesticide exposure from agricultural activities and former agrochemical factory in a Salvadoran rural community. Environmental Science and Pollution Research, 24(2), 1662–1676.

[23]Azaroff, L. S., & Neas, L. M. (1999). Acute health effects associated with nonoccupational pesticide exposure in rural El Salvador. Environmental research, 80(2), 158–164.

[24]de Campos, M., & Olszyna-Marzys, A. E. (1979). Contamination of human milk with chlorinated pesticides in Guatemala and in El Salvador. Archives of environmental contamination and toxicology, 8(1), 4.

[25]López, A., Ribó, A., Quinteros, E., Mejía, R., Alfaro, D., Beltetón, W., … & López, D. L. (2015). Riesgo de Exposición a Contaminantes Nefrotóxicos en Las Comunidades Las Brisas. El Salvador, 1–15.

[26]Murray, D. L. (1994). Cultivating crisis: the human cost of pesticides in Latin America. University of Texas Press. Austin (USA).

[27]See Footnotes 13 and 16.

the period of the civil war, the cultivation of cotton was in the lowlands progressively replaced by that of sugarcane. In El Salvador, as in other neighboring countries, this monoculture has a progressive growth until today.[28] The cultivation of sugarcane also involves the massive use of agrochemicals. Apart from pesticides and fertilizers, chemical ripeners, such as glyphosate, are massively applied aerially.[29] Finally, in Central America the traditional way of harvesting sugarcane, known as zafra, consists of burning the sugarcane to remove the foliage and manually cutting it with a machete. The communities neighboring the sugarcane fields are exposed to the agrochemicals, which are massively applied by airplane, as well as to particles transported by the smoke produced by the burning of sugarcane. Additionally, the cane cutters are subjected to extreme working conditions that involve ambient temperatures above 40 °C, the effort of manual cutting, the cutting edges of the plants and the absorption of particles produced by burning. Peraza et al. have identified a higher prevalence of CKDu among sugarcane workers and former cotton workers in the region.[30] Agrochemicals were also applied to extensive crops like coffee, corn, beans and other vegetables. A High CKDu prevalence was also found in two Salvadorian farming communities, in which maize and beans were the main cultivated crops.[31]

But also the geological characteristics of El Salvador play an important role in investigating the reasons for the high CKDu prevalence.

Fluxes of volcanic gases pollute Lago Ilopango waters (one of the largest fresh water reservoirs in El Salvador) with high levels of boron (B) and arsenic (As) and make the reservoir unsuitable for human water consumption.[32] A similar pollution was also found in other lakes in El Salvador.[33] Anticó, Cot, Ribó, Rodríguez-Roda and Fontàs analyzed the intake water (obtained mainly form wells and ponds) from Torola, a village located in Morazán highlands in El Salvador. The intake water was

[28]Food and Agriculture Organization of the United Nations (FAO) (2019, June, 13): *Production of Crops, Area harvested of Sugarcane in El Salvador (1961–2017)* Retrieved from http://www.fao.org/faostat/en/#data/QC.

[29]CENGICAÑA (Centro Guatemalteco de Investigación y Capacitación de la Caña de Azúcar). (2014). *El Cultivo de la Caña de Azúcar en Guatemala.* Melgar, M.; Meneses, A.; Orozco, H.; Pérez, O.; y Espinosa, R. (eds.). Guatemala. 512 p.

[30]Peraza, S., Wesseling, C., Aragon, A., Leiva, R., García-Trabanino, R. A., Torres, C., … & Hogstedt, C. (2012). Decreased kidney function among agricultural workers in El Salvador. *American Journal of Kidney Diseases, 59*(4), 531–540.

[31]Vela Parada, X. F., Henríquez Ticas, D. O., Zelaya Quezada, S. M., Granados Castro, D. V., Cuchillas, H., Marcelo, X., & Orantes Navarro, C. M. (2014). Chronic kidney disease and associated risk factors in two Salvadoran farming communities, 2012. *MEDICC review, 16*(2), 55–60.

[32]López, D., Ramson, L., Monterrosa, J., Soriano, T., Barahona, J., & Bundschuh, J. (2009). Volcanic arsenic and boron pollution of Ilopango lake, El Salvador. Natural arsenic in groundwater of Latin America. En: J. Bundschuh and P. Bhattacharya (series eds): Arsenic in the environment, 1, 129–143.

[33]Cabassi, J., Capecchiacci, F., Magi, F., Vaselli, O., Tassi, F., Montalvo, F., … & Caprai, A. (2019). Water and dissolved gas geochemistry at Coatepeque, Ilopango and Chanmico volcanic lakes (El Salvador, Central America). *Journal of Volcanology and Geothermal Research, 378,* 1–15.

polluted with aluminum (Al) and nickel (Ni) levels above the Salvadoran guidelines for drinking water safety.[34]

Metal mining is a common source of pollution in Central America. Although the Salvadoran government prohibited metal mining in 2017, pollution caused by mining is still one of the main problems: pollution is caused by former mines or by illegal handmade metal mines. Moreover, mines located in neighboring countries but discharging pollutants into tributary basins of El Salvador cause pollution of rivers and groundwater, too.[35] Illegal waste dumps and uncontrolled residual water discharges from industrial and urban areas directly into rivers and lakes are also sources of heavy metal pollution in El Salvador.[36]

2 Technology and Modelling Aspects Related to CKD

2.1 Space Technology and Mitigating the Risk Caused by CKDu

As mentioned in the previous chapter, the etiology of CKDu is not fully clarified. The considered chemicals like some types of pesticides[37] as well as boron[38] and arsenic[39] are considered as nephrotoxic. A lot of studies show, that CKDu occurs in El Salvador mostly in regions with high agrochemical usage and in regions, where the water is contaminated with heavy metals. Even if there is no direct proof for the responsibility of agrochemicals and heavy metals for CKDu, these substances can have negative impacts on human and environmental health. Reducing the application and the exposure to these chemicals to the only necessary amount has positive effects on human and ecosystem health and might lead to an economic benefit of farmers. The focus on risk mitigation strategies related to agrochemicals and heavy metals addresses economic benefits (e.g. by obtaining the same harvest yield with less agrochemicals), less agrochemicals in environmental and water resources.

---

[34]Anticó, E., Cot, S., Ribó, A., Rodríguez-Roda, I., & Fontàs, C. (2017). Survey of Heavy Metal Contamination in Water Sources in the Municipality of Torola, El Salvador, through In Situ Sorbent Extraction. *Water, 9*(11), 877.

[35]Garay Zarraga, A. (2014). La minería transnacional en Centroamérica: lógicas regionales e impactos transfronterizos. El caso de la mina Cerro Blanco.

[36]Rosales-Ayala, F., & Campos-Rodríguez, R. (2019). Gestión de las aguas residuales en la ciudad La Libertad, El Salvador. *Revista Tecnología en Marcha, 43.*

[37]Ghosh, R., Siddarth, M., Singh, N., Tyagi, V., Kare, P. K., Banerjee, B. D., … & Tripathi, A. K. (2017). Organochlorine pesticide level in patients with chronic kidney disease of unknown etiology and its association with renal function. *Environmental health and preventive medicine, 22* (1), 49.

[38]Pahl, M. V., Culver, B. D., & Vaziri, N. D. (2005). Boron and the kidney. *Journal of renal nutrition, 15*(4), 362–370.

[39]Singh, A. P., Goel, R. K., & Kaur, T. (2011). Mechanisms pertaining to arsenic toxicity. *Toxicology international, 18*(2), 87.

In the following chapter, we will demonstrate some risk mitigation strategies, whereby space technology is considered to be one element in a multidisciplinary risk mitigation approach for CKDu. In this sense, space technology is used to

- determine the geolocation of risk exposed people
- create maps, required to mitigate the risk
- deliver the best fitting risk mitigation strategy to the user.

2.2 Spatial Decision Support System (SDSS)

Especially in rural areas in developing countries, people do not have the information about existing and available risk mitigation strategies. Furthermore, not every risk mitigation strategy is applicable for every person due to local and regional requirements and constraints. To increase the effectiveness of a risk mitigation regime, a Spatial Decision Support System (SDSS) can be used. In the following chapters we will discuss the use of a SDSS with the aim to mitigate the risk caused by agrochemicals or by heavy metals.

The fitness of a risk mitigation strategy is inter alia dependent on personal parameters (like literacy, involvement in the agricultural working process, health history) and on parameters with a temporal and spatial dimension, like the availability of resources, actual pesticide concentrations in the environment etc. By incorporating these parameters in the decision support process, the best fitting risk mitigation strategy, tailored to each person, should be proposed by the SDSS. In other words, a SDSS for our task helps to find the most effective decision path.

A SDSS is defined as a computer based system which should help to solve a spatial problem. In general, a SDSS consists of two components, a Decision Support System (DSS) and a Geographic Information System (GIS).[40]

To handle temporal and spatial data a geodatabase can be used and connected to the GIS. By the knowledge of the GPS (Global Positioning System) position of a SDSS user, information handled by the GIS and stored in temporal and spatial data maps or databases can be determined with the aim to use them for the decision making process. In a DSS models for the decision making process are implemented. There are different methodical principles available, which can be implemented in a SDSS, like fuzzy logic[41] or the principles of neural networks.[42]

---

[40]Keenan, P. B. (2003). Spatial decision support systems. In Decision-Making Support Systems: Achievements and Challenges for the New Decade (pp. 28–39). IGI Global.

[41]Makropoulos, C. K., Butler, D., & Maksimovic, C. (2003). Fuzzy logic spatial decision support system for urban water management. Journal of water resources planning and management, 129 (1), 69–77.

[42]Shim, K. C., Fontane, D. G., & Labadie, J. W. (2002). Spatial decision support system for integrated river basin flood control. Journal of Water Resources Planning and Management, 128 (3), 190–201.

We suggest to combine the models of fuzzy logic and neural network, whereby the advantages of both techniques can be combined: a system based on logical *"if ... then ... "* implications and at the same time with a learning ability.[43] The ability to learn addresses the requirement to integrate medical and environmental monitoring data as input data and new scientific results as rules in a SDSS. The grade of fitness is assigned to each available risk mitigation strategy. The grade of fitness is determined by the SDSS and the risk mitigation strategy with the highest fitness is then proposed to the user or a selection of available risk mitigation strategies with the calculated grades of fitness is offered to the user to choose from. Space technology is needed to obtain the actual geolocation of a user. The SDSS can evaluate the location of the user with the fitness grade of a risk mitigation strategy in the GIS, e.g. the availability of a health service delivery in the area of the user. The SDSS can learn from the feedback of the user, e.g. by a subjective rating of the user about the fitness of the proposed risk mitigation strategy or just by the selection of the provided options for risk mitigation strategies with the assigned grades of fitness. Users might select risk mitigation strategies that have not the highest rated grades of fitness by the SDSS. So, the SDSS becomes also an analytic tool for user preferences. This ability to learn is integrated in the approach, so that user preferences or local requirements and constraints can be integrated in the decision support and the quality of the decision making process is increased and tailored to the geolocation of the user.

Such a SDSS can be for example implemented as an application for a mobile device, whereby the SDSS user is supported by the application. In addition to the technology level of the SDSS reaching the risk exposed people with services and local and regional capacity building strategies is the key for the implementation of available and accepted risk mitigation strategies for the risk exposed communities.

2.3 Low-Cost NDVI Technique

The Normalized Difference Vegetation Index (NDVI) is a vegetation index, which can be determined with the help of multi-spectral images of the earth surface. To calculate the NDVI the reflection rates of red and near infrared light are used. With the help of the NDVI, green and healthy vegetation can be distinguished from less green and unhealthy vegetation, as healthy vegetation has different reflection rates than unhealthy vegetation.[44] Based on the NDVI and other assessment methods for the crop, e.g. by excluding other factors being responsible for the unhealthy condition of the crops than pests or poor nutrition, tailored application of agrochemicals can be implemented.[45,46]

[43]Borgelt, C., Klawonn, F., Kruse, R., & Nauck, D. (2003). Neuro-fuzzy-systeme: Von den Grundlagen künstlicher neuronaler Netze zur Kopplung mit Fuzzy-systemen. Wiesbaden: Vieweg.
[44]Pettorelli, N. (2013). The normalized difference vegetation index. Oxford University Press.
[45]Schumann, A. W. (2010). Precise placement and variable rate fertilizer application technologies for horticultural crops. HortTechnology, 20(1), 34–40.
[46]Elliott, N. C., Backoulou, G. F., Brewer, M. J., & Giles, K. L. (2015). NDVI to detect sugarcane aphid injury to grain sorghum. *Journal of economic entomology*, 108(3), 1452–1455.

The use of NDVI and precision farming is in general implemented in highly technically equipped farming infrastructures. The used equipment in industrialized agriculture is mostly not affordable in developing countries.[47] Low-cost precision farming applies these techniques to environments with mobile devices (smart-phones) and local and regionally applied farming infrastructure (backpack sprayers).

A possible approach for low-cost precision farming would be the use of free available multi-spectral satellite images, e.g. retrieved from ESA's Copernicus Programme[48] or from Landsat satellite data,[49] to produce NDVI maps. The multi-spectral images can be processed with an open source GIS. For example the software tool GRASS GIS has an implemented command for the creation of NDVI maps.[50] The NDVI maps can be connected to a SDSS.[51] The user can be informed about areas and regions, where the crop health is not optimal, the farmer can inspect these areas and adjust the fertilizer or pesticide amount to the needs of the crops. This can lead to a decreasing amount of used agrochemicals and contributes to a reduction of the exposition to these substances. However, the spatial resolution of free available satellite images and low cost NDVI maps are not as high as produced by a commercial multi-spectral sensor attached to a tractor in commercial agriculture. Remote sensing technology can be regarded as one methodology among others to reduce the application of agrochemicals together with ground based assessment methods.

Reducing the application of partly toxic agrochemicals has not only the advantage of less exposition of farm workers and other people involved in the agricultural working process, but the strategy provides also an economic benefit. If the main driver for the application of low-cost precision farming techniques are economic reasons, then the application should provide an economic benefit. In general, precision farming reduces the amount of required agrochemicals with the same harvest yield. This economic benefit could have environmental health and public health benefits, too, if agrochemicals are (nephro) toxic or have a negative impact on biodiversity, contamination of water resources, etc.

[47]Grisso, R. D., Alley, M. M., Thomason, W. E., Holshouser, D. L., & Roberson, G. T. (2011). Precision farming tools: variable-rate application.

[48]European Space Agency (ESA) (2019, May, 25): Copernicus observing the earth. Retrieved from https://www.esa.int/Our_Activities/Observing_the_Earth/Copernicus/Overview3.

[49]United States Geological Survey (2019, June, 06): Landsat Missions. Retrieved from https://www.usgs.gov/land-resources/nli/landsat.

[50]GRASS GIS (2019, May, 25): Documentation. Retrieved from https://grass.osgeo.org/documentation/.

[51]Zhang, X., Shi, L., Jia, X., Seielstad, G., & Helgason, C. (2010). Zone mapping application for precision-farming: a decision support tool for variable rate application. Precision agriculture, 11 (2), 103–114.

2.4 Open Educational Resources

The implementation of a SDSS requires capacity building and learning. The risk mitigation strategies and precautionary principles for risk mitigation can be then used by the local communities. The different local requirements and constraints need the adaption of educational resources to these requirements and constraints.

In this context we use the term Open Educational Resources (OER) as defined by the UNESCO in 2002 at the Forum on Open Courseware. It defines OER as "teaching, learning and research materials in any medium, digital or otherwise, that reside in the public domain or have been released under an open license that permits no-cost access, use, adaptation and redistribution by others with no or limited restrictions. Open licensing is built within the existing framework of intellectual property rights as defined by relevant international conventions and respects the authorship of the work".[52] As mentioned above, we use OER in a geospatial context as Spatial Open Educational Resources (SOER) by assigning the teaching, learning, and research resources to one or more geolocations in which the OER is applicable. It means that the OER are tailored to the local and regional constraints. The requirement that OER have been released under an intellectual property license that permits their free use and re-purposing by others, allows the adaptation of OERs to a specific geolocation or area of interest, to maximize the local acceptance of full courses, course materials, modules, textbooks, streaming videos, tests or risk mitigation strategies in general.[53]

In this context, SOERs can be used e.g. as a manual for the proper use of pesticides or for the use of low cost risk mitigation strategies, as described in the previous chapter. Also videos, pictures and other documents, which help to increase the risk literacy, can be developed as OER. However, not every OER fits to each user, as there might be language, social or cultural obstacles. Therefore, the OER have to be provided in different local languages, with different local availability of resources and for different user groups.

The localization of learning resources can be implemented by OER authors with the adaption of learning resources to requirements and constraints at the geolocation (*lat, long*) of the user.

Again, in this approach space technology is used to determine the user's location. If users access the OER with mobile devices the GPS sensor can be used to tailor the OER for the requested geolocation, if there are OER available for that purpose. The requirement for this is an existing IT-infrastructure.

---

[52]UNESCO (2012). World open educational resources congress. Retrieved from http://www.unesco.org/new/en/communication-and-information/events/calendar-of-events/events-websites/world-open-educational-resources-congress/.

[53]Peña-López, I. (2007). Giving knowledge for free: The emergence of open educational resources. "Giving Knowledge for Free: The Emergence of Open Educational Resources" (2007) (PDF: http://www.oecd.org/education/ceri/38654317.pdf). Center for Educational Research and Innovation. Retrieved 28 May 2019.

GIS tailored OER can be adapted to the geolocation[54] by

• adapting the language of the OER (translation) for the citizens that want to access the capacity building material for risk mitigation strategies (simple language with local examples or references to local or regional terminology and wording)
• using local examples that match local social, cultural constraints and integrate available resources for a specific region or community.

Beside the spatial evaluation of OERs additional aspects can be integrated into the adaption of OER to the mobile device users, e.g. by adapting the provided capacity building materials to the profile and skills of a mobile device user (e.g. a risk exposed farm worker gets different OER in comparison to public health worker). Furthermore, an institutional localization can be applied, because different organizations have a different focus and management for risk mitigation and must operate in different institutional or organizational regulations and within regional requirements and constraints.

An example, how documents can be used to mitigate the risk caused by agro-chemicals, is given in the publication from Ribó, Mejía, Quinteros and López.[55] In this publication a basic description of good pesticide practice in each of several stages of pesticide handling (acquisition, transportation, storage, formulation, application and waste management) is described. It is like a handbook orientated to health agents in rural areas, partly written, partly with easy understandable images. The publication focuses on people with basic literacy and with Spanish as language and helps people to increase their risk literacy and to reduce the exposition to pesticides even if they use the same amount of agrochemicals as before. This publication can now be assigned with a geolocation, e.g. the locations with Spanish as mother language.

The publication was not released under an open license, therefore it cannot be considered as an OER. That means, if we want to translate the document in a different language for other countries affected by CKDu permission is needed. Even for minor adaptations to local requirements and constraints an extra permission to remix and alter the source is needed. Furthermore, an automated adaption of SOER to the local requirements and constraints by the SDSS is not allowed.

If the owner does not want to release the document in a different language it would not be possible. Under an open license, it would be no problem to translate the document into another language or to transfer the written words into images in a way, that also people with a low literacy can understand the message (see licensing model of Creative Commons https://creativecommons.org/choose/). If the original publication would be translated into a different language, like Sinhala as spoken in

[54]Belgiu, M., Strobl, J., & Wallentin, G. (2015). Open geospatial education. ISPRS International Journal of Geo-Information, 4(2), 697–710. (https://www.mdpi.com/2220-9964/4/2/697).
[55]Ribó A., Mejía R., Quinteros E. & López A. (2015) Manejo de Plaguicidas en la Agricultura. 1st ed. San Salvador, El Salvador: Instituto Nacional de Salud, Ministerio de El Salvador; 2015. 50 p.

Sri Lanka, the translated publication would get another geolocation, in this case all locations, where Sinhala is spoken.

By determining the actual location of a risk exposed person via space technology and GPS, it would be possible to propose the best fitting OER to the user.

2.5 Spatial Modelling of Risks

The knowledge about the spatial distribution of risk (risk maps) is essential to initialize appropriate risk mitigation strategies. The grade of risk a person is exposed to can be determined by comparing the actual location of a risk exposed person, determined via space technology (GPS), and the grade of risk stored in a risk map at the actual geolocation of the person. The spatial distribution of the grade of risk is mostly stored and visualized in risk maps. In this section, we will demonstrate some principles of generating data showing the spatial distribution of risk, related to the case of CKDu in El Salvador.

2.5.1 Fuzzification of Thresholds for Agrochemicals

In the risk assessment, the basic principle to analyze the risk of a substance in the environment is to compare the measured or modeled concentration c of the substance in the environment with a regulatory threshold value r for the regarded chemical. If the concentration c in the environment exceeds the threshold value r, then it is assumed, that there exists a risk.

The Normalized Threshold Index (NTI) uses the regulatory threshold r provided by a national environmental agency or ministry for normalization of the measured concentration c. The fraction

$$n := \frac{c}{s}$$

indicates if the measure concentration is exceeding the regulatory threshold (i.e. $n > 1$) or if the measured concentration c is below the regulatory threshold (i.e. $n < 1$). Furthermore, the NTI shows the factor in which the concentration is exceeding the regulatory threshold (i.e. $n = 5$ means that the measure concentration is 5 times higher than the regulatory threshold). In this example, we have a crisp value to determine if there is a risk at a location ($n > = 1$) or if there is no risk ($n < 1$). Let's assume, that we have a regulatory threshold value of a substance of 100. An environmental concentration of 100.1 would mean, that we have a risk at the given location, whereby a concentration of 99.9 would mean, that at that location there is no risk, beside the fact, that the difference between the concentrations at the locations is only little. Therefore, we will demonstrate an approach called fuzzyfication, how such crisp changes between "risk" and "no risk" can be avoided by using the NTI.

Fuzzyfication in this sense means, that we assign a value between 0 and 1 to each geolocation, whereby 1 means, that at the current geolocation we have a high risk, 0 means there's no risk and 0.5 means, that at the location a risk of degree 0.5 can be observed.

For the fuzzification of the NTI a sigmoid function

$$sig_{s,d}(x) := \frac{1}{1 + e^{-\frac{x-s}{d}}}$$

is used. The positive value $d > 0$ has an influence on the steepness of the sigmoid function at the threshold s. Small values of d result in an increase from $sig_{s,d}(x)$-values close to 0 to $sig_{s,d}(x)$-values close to 1 in a small range of x-values, while large values of d define an increase of $sig_{s,d}(x)$ from close to 0 to close to 1 in a longer range of x-values (less steep—smaller maximal value of the deviation $sig_{s,d'}(x)$ The sigmoid function can be used for decision making. Values close to one mean high risk and need monitoring activities to protect the population or create early warning tailored to the geolocation of the user. Assume the environmental agency or public health agencies have a national grid of a monitoring system. The geolocations of the measurements are available in a GIS. By the application of the NTI on a spatial dimension we obtain a Spatial Normalized Threshold Index (SNTI) that can be displayed like a Digital Elevation Modell (DEM) in a map. The SNTI can be regarded as a risk map for the concentration c (lat, $long$) at the geolocation (lat, $long$).

This spatial approach can be applied to the fuzzification of the NTI to the SNTI by a function called

$$sig_{map}(lat, long) := sig_{s,d}(c(lat, long))$$

i.e. the concentration $c(lat, long)$ at geolocation (lat, $long$) is used as x-value for the sigmoid function $sig_{s,d}$. The function sig_{map} creates a fuzzy layer in a GIS that displays areas with a higher demand for monitoring. The alpha-cut of sig_{map} with $sig_{map}(lat, long) > \alpha$ (e.g. with $\alpha = 0.9$) displays areas on the map, in which the monitoring activities in future are recommended. An α-value closer to 1 reduces the area for further monitoring activities and limits the monitoring activities to areas with higher concentration.

2.5.2 Spatial Information Density

Assume we define a Spatial Information Density (SID) by the number of measurements per $100\ km^2$. Let A be the considered area with m points of measurements. The two-dimensional integral of the non-negative SID over the area A provides as result m, the basic definition of SID is a density function (non-negative, integrable) with the domain of geolocations (lat, $long$).

SID $(lat, long) = 0.5$ indicates that the average information density at geolocation $(lat, long)$ is 0.5 measurements per 100 km^2, a $SID(lat, long) = 2$ indicates that at geolocation $(lat, long)$ we have 2 measurements per 100 km^2. With

$$dens_{x_k, y_k}(lat, long) := \frac{m_k}{1 + \frac{1}{d} \cdot (\textbf{lat} - x_k)^2 + \frac{1}{d} \cdot (\textbf{long} - y_k)^2},$$

we can calculate a value m_k at the geolocation (x_k, y_k), meaning the density for a single geolocation (x_k, y_k) of a measurement has the maximum value m_k at the geolocation (x_k, y_k) and the value is decreasing dependent on the distance to the location (x_k, y_k). We define a density function

$$D(lat, long) := \sum_{k=1}^{k_{max}} dens_{x_k, y_k}$$

with k_{max} geolocations and $m_1, \ldots, m_{k_{max}}$ measurements at each geolocation. Each density function $dens_{x_k, y_k}$ is integrable.

$m_{max} := m_1 + \ldots + m_{k_{max}}$ defines the number of all monitoring data in the area A. The integral

$$\int_A D(x, y)dxdy = a$$

is used to normalize the density function by

$$SID(lat, long) := \frac{m_{max} \cdot D(lat, long)}{a}.$$

The SID can now be used for measurable subsets A_0 of A to determine the information density for A_0. The average information density for A_0 can be calculated by the integral

$$\int_{A_0} SID(x, y)dxdy$$

divided by the size of the area of A_0.

For decision makers it is important to know the SID at a specific geolocation because the allocation of resources can be costly and a low $SID(lat, long)$ at the geolocation might lead to different decisions. E.g. a high $SNTI(lat, long)$ and high $SID(lat, long)$ lead to the implementation of risk mitigation strategies at geolocation $(lat, long)$, while a high $SNTI(lat, long)$ and a low $SID(lat, long)$ lead to an increasing monitoring activity in a specific region to have a better availability of data in that region for the decision making processes.

The SID can be provided with GPS-sensors on mobile devices for the geolocation of the risk exposed citizen or for the public health worker implementing risk mitigation strategies.

3 Conclusions for Next Steps

The consideration of One Health risks (i.e. environmental health + animal health + public health) requires an integrated risk analysis and an assessment of available response activities according to risk. In this publication, we addressed the contribution of space technology to mitigate the risk for CKDu.

The application of space technology in the domain can be driven by push factors from space technology and by pull factors from the health domain. Both factors have some short comings. First, the health domain might not know about the requirements and constraints of space technologies and the possible services. Secondly, space technology might not be aware of the requirements and constraints of the health domain and therefore the space technology might push services in the health domain that might not be applicable in the current form or do not match the requirements and constraints of the health domain. For example, space technology is able to detect light emission during night, but these capabilities do not give any direct link towards the health domain. But if we add the information that light emission is an indicator for the availability of electricity (pre-post-disaster comparison) and the availability of electricity is crucial for providing health care services in a hospital or in a rural health care facility, the link between the application of space technology in the context of disaster management is visible. The next step in decision making is important on a risk management level. Space technology is regarded as one option/methodology among others in the definition of risk mitigation strategies in the health domain. The investment in a specific methodology is compared with the prospective impact in the health domain. If we apply that to CKDu the response in the area with health care services and continuous monitoring of kidney functions can be extrapolated to regions with less monitoring data and the same exposure pattern and toxins in the environment and water reservoirs. Furthermore, medical services e.g. monitoring of kidney health, options for dialysis or in the final stage of kidney damage kidney transplant, can be mapped spatially according to the availability and the access to patients and risk exposed communities. Leaving the area of health care interventions, the response in the area of environmental sciences and agro-science is the reduction of exposure to nephrotoxic substances. Spatial patterns of risk can be identified by the characteristics of the workflow with agrochemicals and the amount of toxicants used by communities. The geospatial analysis of layers stored in a GIS is the foundation of the decision support. Based on scientific evidences and with the application of precautionary principles multi-factorial risk factors can be considered and incorporated in the decision making process.

In general space technology can be involved by the application of satellite communication, navigation with satellite navigation system and/or remote sensing and detection of weather conditions, crop health, etc. The main starting point is to use the GPS sensor on mobile devices to provide tailored risk mitigation information for the current geolocation of the user. This approach can be extended to low cost precision farming to reduce e.g. the number of agro chemicals or to gain an economic benefit by getting the same harvest yield with the application of less agrochemicals. Also OER can be assigned to specific geolocation. So the described risk mitigation resources are applicable at a specific geolocation and the language and examples used in the OER content use local and regional examples.

Jörg Rapp studied environmental sciences and is a PhD candidate and research associate in the working group for Mathematical Modeling at the Institute of Mathematics of the University of Koblenz-Landau, Campus Landau. His research focuses on the interface between applied mathematics and environmental science.

Prof. Engelbert Niehaus is head of the working group for Mathematical Modeling at the Institute of Mathematics and head of the Computer Science Centre of the University Koblenz-Landau, Campus Landau. His research interests include the use of open educational resources in educational processes as well as the development of mathematical methods that can be used in the area of risk mitigation. He is a member of the United Nations OOSA Expert Focus Group Expert Committee on Space and Global Health, where he is responsible for mathematical modeling in the development of an "epidemiological early warning system".

Dr. Alexandre Ribó is a geologist experienced in geological risks, GIS and International Development Cooperation. He worked for six years in El Salvador, first for World Geologists, a NGO with the focus on developing geological risk mitigation strategies at the community level, and then as a researcher at the Instituto Nacional de Salud of El Salvador in topics related to environmental health and GIS. He focused his research on environmental risk factors related to CKDu. Nowadays he lives in Barcelona working as a geologist in the private sector and as an external GIS consultant for the UN.

Roberto Mejía holds a bachelor's degree in Environmental Health. Since 2011 he has worked in the field of environmental research at the National Institute of Health of El Salvador. He has published different articles related to public health and GIS.

Edgar Quinteros holds a bachelor's degree in environmental health and a master's degree in epidemiology. He has five years of working experience in the field of environmental toxicological studies. Currently, he works at the National Institute of Health of El Salvador and is a member of the editorial committee of Alerta, a scientific journal.

Anna Fath studied mathematics, physics and educational sciences and is part of the working group for Mathematical Modeling at the Institute of Mathematics of the University of Koblenz-Landau, Campus Landau. Her research interests are in the area of simulating and visualizing flooding events.

Kourou: The European Spaceport and Its Impact on the French Guyana Economy

Anne-Sophie Martin

Abstract

Space programs such as telecommunications, remote sensing and navigation, are fundamental activities for the development of the Latin American countries, and their interests in these fields are growing for several years now. However, they are still bound to require some external assistance related to satellite construction and launching, as well as for accessing to advanced technologies. This contribution focuses on the European spaceport, Kourou, in French Guyana, and on how it contributes to the country's growth. Firstly, it analyzes the legal regime of the Center. Then, the contribution addresses the aspect of international cooperation with the coming of new launchers inside the spaceport. Finally, it discusses the importance of Kourou for French Guyana. Indeed, the role of the *Centre National d'Etudes Spatiales* through various initiatives with Universities and space industries are particularly important for the economic development and the educational plan of the country.

A.-S. Martin (✉)
Department of Political Sciences, Sapienza University of Rome, Rome, Italy
e-mail: annesophie.martin@uniroma1.it

© Springer Nature Switzerland AG 2020
A. Froehlich (ed.), *Space Fostering Latin American Societies*, Southern Space Studies, https://doi.org/10.1007/978-3-030-38912-3_6

1 Introduction

Latin America's space programs have continued to develop and expand in recent years. While there have been some failures, regional space programs have in general continued to grow, with additional launches planned for the near future.[1] They have enjoyed significant accomplishments as countries like Bolivia and Peru have modern new satellites in orbit that support telecommunications and surveillance projects, while Argentina, Colombia, Ecuador, and Mexico have domestically constructed their own platforms.[2]

This contribution focuses in particular on the French Guyana, and on the impact of the space sector on its economy and its development.

First of all, from an historical perspective, the choice of the French Guyana for the European Spaceport is not due to chance. Among the fourteen sites studied, the country was the best area that met all the criteria for launch activities.[3]

After the independence of Algeria in 1962, the *Centre National d'Etudes Spatiales* (CNES) started to seek a new base near Equator, an area that would allow carrying out all space missions in the best conditions. The Ground Equipment Division of CNES's Scientific and Technical Department studied various possibilities. Indeed, diverse criteria had to take into account,[4] such as the possibility of polar and equatorial launches, proximity of the Equator, sufficient dimensions to ensure the safety of launches, existence of a deep-water port with appropriate means of handling, as well as political stability.

Finally, on April 14, 1964, the Prime Minister, Georges Pompidou, chose Guyana that presented many appropriate features[5]: a wide opening on the Atlantic Ocean favoring all launching missions both for geostationary orbit and for polar orbit with minimal risk to the population and the surrounding property; the proximity of the Equator allowing, in case of geostationary orbit transfer, to make just a few changes to a satellite's trajectory, and enabling launchers to profit from the "slingshot" effect, that is the energy provided by the speed of the Earth's rotation around the axis of the Poles[6]; the low density of the population; the possibility to install tracking systems on the surrounding hills (radars and telemetry antennas); good weather; existing infrastructures relatively simple to adapt to the needs of the future space center (roads, aerodrome, ports, telecommunications, etc.). In this context, the Guyana Space Center (GSC) moved to Kourou in 1965.

---

[1]Sanchez (W. A.), *Latin America's Space Programs: An Update*, The Space Review, January 22, 2018: http://www.thespacereview.com/article/3413/1 (last accessed May 24, 2019).
[2]Ibidem.
[3]*Centre Spatial Guyanais*: http://www.cnes-csg.fr/web/CNES-CSG-fr/9777-implantation.php (last accessed May 24, 2019).
[4]Ibidem.
[5]Ibidem.
[6]European Space Agency website, *Europe's Spaceport*: http://www.esa.int/Enabling_Support/Space_Transportation/Europe_s_Spaceport/Europe_s_Spaceport2

CNES exercises overall management of the GSC in several ways. First of all, the land on which the GSC is located belongs to CNES. Then, the installations and ground segments, as well as the planning and supervision of the operations necessary for the launches are part of its skills. Finally, the president of the CNES exercises, on behalf of the State, the special police on the exploitation of the installations.

Its missions are in line with the objectives pursued by the European Union in the field of space transport, namely to ensure an independent access to space in order to enable it to carry out the missions which come under its competences. To achieve this objective, France and the other European Space Agency (ESA) Member States have recognized the strategic importance of own launching facilities and decided in 1973 to build them at GSC in order to hold a leading position in the launch services market and maintain the expansion and expertise of its industry. In this context, it was decided in 1980 to create the company Arianespace, commercial operator of European launchers Ariane and responsible for launch services.[7]

Since the decision to locate the Kourou space base in 1965, history has been marked by differentiated periods in terms of investment, commercial success and technological progress. The impact of space development, combined with the migration periods seen in Guyana, have shaped the economic development of the territory.

The establishment of the Space Center is accompanied by massive investments for the construction of infrastructures and the new city of Kourou with modern districts, medical center and hotels. The territory of Guyana has also been marked by the infrastructures built during this period: improvement of the Cayenne-Kourou road, bridge over Kourou, port of Pariacabo, along the runway of the airport. Guyana's GDP, which was very weak, has begun a period of sustainable growth.

The technical and commercial successes of Ariane 5 make it possible to increase the activity of the Center with the diversification of the range of launchers.[8] This diversification became reality from 2011 with the first launch of Soyuz, and in 2012 with Vega.[9]

The Ariane 6 project opens a new period with important investments. The overall development cost of the new launch facility (ELA 4) for the future launcher is estimated at 600 million euros, of which 94 million euros are local contracts (around 15%).[10] Ariane 6 launch vehicle will be capable of a wide range of mission. It will also have the flexibility to launch both heavy and light payloads inot

---

[7]Arianespace website: http://www.arianespace.com/company-milestones/ (last accessed May 24, 2019).
[8]*Declaration by Certain European Governments on the Launchers Exploitation Phase of Ariane, Vega and Soyuz from the Guiana Space Centre*, Paris, 30 March 2007: https://assets.publishing.service.gov.uk/government/uploads/system/uploads/attachment_data/file/238562/7700.pdf (last accessed May 24, 2019).
[9]https://www.universalis.fr/encyclopedie/centre-spatial-guyanais/3-soyouz-et-vega/ (last accessed May 24, 2019).
[10]SpaceNews, *Airbus Safran Agrees to $440 Million Ariane 6 Contributions*, May 29, 2015: https://spacenews.com/airbus-safran-agrees-to-400-million-ariane-6-contribution/ (last accessed May 24, 2019).

orbit. On September 2018, Arianespace signed two contracts for Ariane 6: the first one with Eutelsat; and the second with the CNES and the *Direction Général de l'Armement* (DGA), the French Defense Procurement Agency.[11]

This period is also marked by a significant change in the industrial landscape of the space industry. At the end of June 2016, Airbus and Safran finalized their joint venture, Airbus Safran Launchers (ASL), a new leader in European launchers to develop Ariane 6.[12] After the repurchase of the CNES' shares, ASL owns 74% of Arianespace's capital and is renamed ArianeGroup on July 1st, 2017.[13] The strategic objective is to respond to the SpaceX offensive and to maintain Europe's leadership in a fast-changing commercial launch market.

In this context, the contribution addresses how the Center is oriented towards the future Sect. 2 with an overview of the Kourou's regulation system Sect. 2.1. Moreover, the contribution focuses on the importance of the cooperation in the field of launchers and the access to outer space Sect. 2.2. Finally, it analyzes the impact of the space sector on the French Guyana economy Sect. 3.

2 Guyana Space Center: A Spaceport Sets for the Future

2.1 An Overview of the Center's Regulation System

Until 2008, the GSC was governed by three intergovernmental agreements.[14]

Firstly, the agreement on the Guyana Space Center[15] concluded on April 11, 2002 which concerned the launching support facilities. It was initially concluded for a period of four years. Through this agreement, France renewed the guarantee of availability, access priority use of CNES facilities and means to the GSC to the ESA and its Member States. Among the main provisions, it was foreseen that the CNES was the landowner of the entire space site, the authority for the design of ground assets made on the site, and the owner of all the technical means implemented for the accomplishment of the missions except for the preparation of the payloads (*EPCU—Ensemble de Préparation des Charges Utiles*) and downstream

---

[11]Arianespace website, *Ariane 6 accelerates as Arianespace signs first commercial GEO multiple-launch contract, plus a new institutional mission*, September 10, 2018: https://www.arianespace.com/press-release/ariane-6-accelerates-as-arianespace-signs-first-commercial-geo-multiple-launch-contract-plus-a-new-institutional-mission/.

[12]Safran website: https://www.safran-group.com/media/airbus-safran-launchers-become-arianegroup-20170517 (last accessed May 24, 2019).

[13]ArianeGroup website: https://www.ariane.group/en/news/airbus-saran-launchers-becomes-74-percent-shareholder-arianespace/ (last accessed May 24, 2019).

[14]French Senate Report n° 308, 2013, p. 8 et ss: https://www.senat.fr/rap/l12-308/l12-3081.pdf (last accessed May 24, 2019).

[15]European Space Agency website: https://www.esa.int/About_Us/Welcome_to_ESA/France_and_ESA_sign_CSG_spaceport_agreement (last accessed May 24, 2019); see also News from Europe's Spaceport, ESA bulletin 112, November 2002, p. 73: http://www.esa.int/esapub/bulletin/bullet112/chapter8_bul112.pdf (last accessed May 24, 2019).

stations. ESA received support from CNES/GSC for Ariane launches in the qualification phase. Availability, freedom of access and use of CNES/GSC facilities were guaranteed for Ariane programs.

Secondly, the Agreement on Launching Assemblies (*Ensemble de Lancements —ELA*) and associated Agency facilities at the GSC, concluded on April 11, 2002. It concerned the means of launching and took into account the evolution of facilities and resources of the ESA on the GSC zone.

Thirdly, the agreement on the Soyuz launchpad, concluded on March 21, 2005 aimed to clarify the conditions under which France authorized ESA to set up, on the GSC site, the Soyuz site and from which all the launches of the Russian launcher from Guyana were carried out, and the conditions of its exploitation. This agreement also specified the regime of international responsibility for these launches, providing that France guaranteed the Agency and its Member States against all claims emanating from a third State or from a national of such State in the event of damage caused during a Soyuz launch operated by Arianespace.

These three agreements have been replaced by a single one taking into consideration primary the operation of the launchers Ariane, Soyuz and Vega from the GSC, and then, the provisions of the French law relating to the Space Operations adopted on June 3, 2008.[16]

This new agreement was signed on December 18, 2008 in Paris.[17] It provides a unified and up-to-date legal basis of the GSC regulation system. It secures the use of GSC facilities by ESA until the end of 2020. The agreement is concluded for a long duration, contrary to the agreement on the previous GSC, successively renewed for a period of four years. Thus, it makes possible to sustain ESA's commitment to financing and using the launching base. Moreover, it provides for

---

[16]*Loi n° 2008-518 du 3 juin 2008 relative aux opérations spatiales*: https://www.legifrance.gouv.fr/affichTexte.do?cidTexte=JORFTEXT000036114413&categorieLien=id (last accessed May 24, 2019). See also *Décret n°2008-1160 du 12 novembre 2008 portant publication de l'accord entre le Gouvernement de la République française et l'Agence spatiale européenne relatif aux ensembles de lancement et aux installations associées de l'Agence au Centre spatial guyanais, signé à Paris le 11 avril 2002*; *Décret n°2009-426 du 16 avril 2009 portant publication de l'accord entre le Gouvernement de la République française et l'Agence spatiale européenne relatif à l'Ensemble de lancement Soyouz (ELS) au Centre spatial guyanais (CSG) et lié à la mise en oeuvre du programme facultatif de l'Agence spatiale européenne intitulé "Soyouz au CSG" et à l'xploitation de Soyouz à partir du CSG, signé à Paris le 21 mars 2005*; *Décret n°2009-1373 du 6 novembre 2009 portant publication du protocole portant amendement de l'accord entre le Gouvernement français et l'Agence spatiale européenne relatif au CSG, signé à Paris le 12 décembre 2006*; *Décret n°2010-375 du 12 avril 2010 portant publication de l'accord entre le Gouvernement de la République française et l'Agence spatiale européenne relatif au Centre spatial guyanais, signé le 11 avril 2002*; *Décret n°2016-1778 du 19 décembre 2016 portant publication de la déclaration de certains gouvernements européens relative àla phase d'exploitation des lanceurs Ariane, Vega et Soyouz au Centre spatial guyanais, adoptéeàParis le30 mars 2007*; *Décret n° 2017-1619 du 27 novembre 2017 portant publication de l'accord entre le Gouvernement de la République française et l'Agence spatiale européenne relatif au centre spatial guyanais et aux prestations associées (ensemble deux annexes), signé à Paris le 18 décembre 2008.*

[17]*ESA and CNES sign contract on Guiana Space Centre (GSC)*: https://www.esa.int/Our_Activities/Space_Transportation/ESA_and_CNES_sign_contract_on_Guiana_Space_Centre_GSC (last accessed May 24, 2019).

all launchers the apportionment of responsibility between the French State and ESA in case of damage caused to third parties.

Comprised of 22 articles, the new agreement consolidates the three former GSC agreements by updating their provisions[18] and by taking into account the above-mentioned French Space Operations Act.

In art. 3, ESA is given the opportunity to consult CNES in order to facilitate the allocation of authorizations or licenses necessary for launches carried out from the GSC. This is a consequence of the law on space operations of June 3, 2008, which states that all launches made from the Guyana Space Center are subject to an authorization regime implemented by the Minister in charge of space activity.

The art. 4 introduces more flexibility in defining the main facilities and means of the GSC launch support package. The list can now be updated by CNES as needed.

The art. 5 replaces the reference to the "GSC safeguard doctrine"[19] with "the GSC safeguarding regulations adopted by CNES", in accordance with the provisions of the Space Operations Act and Decrees. It also adds a safety mission to CNES.

The art. 7 recalls that "the Agency shall comply with the laws and regulations applicable to the GSC with regard to safeguarding and safety".

The art. 8 sets out the practical arrangements for making the GSC available for ESA's activities. The contract concluded between CNES and ESA, establishing the commitments of the first vis-à-vis the second, must now also specify the services to be provided by CNES for the permanent maintenance in operational conditions of GSC. Finally, a new paragraph recalls that Arianespace has committed to cover all costs related to the GSC launch support package allocated to the operation of the Soyuz launcher.

In art. 13, the new wording strengthens legal certainty by providing for the case of "construction of ESA facilities and means outside the lands made available to it under this Agreement", which is the subject of specific agreement between the Parties. ESA has the possibility to do not own the facilities and means built by it on the land placed at its disposal. Protective fences, for their part, are the property of CNES.

The art. 14 introduces a mutual information obligation of the parties in the event that one of them prepares to transfer the facilities and means of the GSC; only the French Government was previously obliged.

---

[18]French Senate Report n° 308, *2013*: https://www.senat.fr/rap/l12-308/l12-3081.pdf (last accessed May 24, 2019).

[19]CSG Safety Regulations, *General Rules*, Vol. 1: http://emits.sso.esa.int/emits-doc/AO-1-5306-AD7-Vol-1.pdf (last accessed May 24, 2019); CSG Safety Regulations, *Specific Rules Spacecraft*, Vol. 2: http://emits.sso.esa.int/emits-doc/ALCATEL/CSG-RS-22A-CNis_05rev_06Vol2Part2_Spacecraft.pdf (last accessed May 24, 2019); *Décret n° 2009-644 du 9 juin 2009 modifiant le décret n° 84-510 du 28 juin 1984 relatif au Centre national d'études spatiales* (safety regulation at the GSC): https://www.legifrance.gouv.fr/affichTexte.do;jsessionid=2472FA1213690B23A9CB503F4B47B259.tplgfr42s_2?cidTexte=JORFTEXT000020719559&dateTexte=20090610 (last accessed May 24, 2019).

The art. 16 incorporates the provisions of the Soyuz GSC Agreement on the registration and jurisdiction of launchers and extends them to Ariane and Vega launchers.

The art. 17 reiterates the established accountability distinction between the GSC agreement and the ELA agreement between the Agency's programs in their development phase and the Arianespace launches in the operating phase.

The art. 20 adds a further clarification to the settlement of disputes, specifying the terms of the vote in the arbitration tribunal, the manner of application of its sentences and the procedure in case of claim.

Finally, the agreement is complemented by two annexes which detail for the first the perimeter of the GSC, as well as the land made available to ESA, and for the second, the main facilities and resources to support launching activities from the GSC.

In addition to provide legal clarifications, this agreement has positive consequences in terms of space strategy. Indeed, the new framework reinforces and perpetuates the independent access of France and Europe to space, guaranteeing long-term cooperation between France and ESA.

Lastly, Arianespace and ESA signed, in 2018, a contractual framework defining procedures for the procurement of launch services by ESA. It applies to all current launchers Ariane 5, Soyuz, Vega, and future Ariane 6, Vega C.[20]

2.2   The Importance of International Cooperation to Further Develop the Launchers Capacity of the GSC

The GSC activities is becoming more significantly[21] with the three launchers Ariane, Soyuz and Vega which offer complementary performances, and allow Europe to launch various types of missions independently.

Ariane rocket is the historic program, launched in 1973.[22] The first launch took place in 1979. Since then, the Ariane 2, 3 and 4 programs have succeeded one another, and now lead to the Ariane 5 launcher. It is a heavy launcher that can launch a load of 9.5 tons in geostationary orbit and nearly 20 tons in low orbit, the largest capacity on the market. Ariane 5 ECA is designed to carry for each launch a large mass satellite (often telecommunications) and a smaller satellite (dual launch principle). A second version exists, Ariane 5 ES, which is mainly intended for Automated Transfer Vehicle (ATV) missions to the International Space Station, the launch of satellite constellations and interplanetary flights.

---

[20]ArianeGroup website: http://www.arianespace.com/press-release/arianespace-signs-frame-contract-with-esa-for-the-procurement-of-launch-services-for-european-space-agency-missions/ (last accessed May 24, 2019).

[21]French Senate Report, op.cit., pp. 7–8: https://www.senat.fr/rap/l12-308/l12-3081.pdf (last accessed May 24, 2019).

[22]Ariane 5, Technical Overview, April 18, 2019: http://www.arianespace.com/wp-content/uploads/2019/04/ARIANESPACE-ENG-FLYER-ARIANE-5-APRIL2019-WEB.pdf (last accessed May 24, 2019).

The Soyuz launcher is an average launcher used for both manned and unmanned flights to the International Space Station (ISS) and for commercial flights under the direction of Starsem. In the Soyuz ST version, the launcher allows to put in orbit low payloads of 4900 kg. This payload can even be 5500 kg in the case of the Soyuz Fregat launcher.[23]

Soyuz at GSC is the emblematic project of Franco-Russian cooperation in the space field. Following an agreement signed between France and the Russian Federation on November 7, 2003, it has been operating at the Guyana Space Center since 2011.[24] It has provided so far 23 missions from the GSC.[25]

A light launcher, Vega designed to complement the family of European launchers in order to meet the market for small missions, including scientific and Earth observation.[26] The program has been funded by seven ESA Member States: Belgium, France, Italy, the Netherlands, Spain, Sweden and Switzerland. It includes the development of the launcher and the construction of dedicated ground facilities at the Guyana Space Center (Vega Launching Site). The Vega launcher is designed to place small payloads (1.5 t) in low orbit. Its inaugural flight took place on February 13, 2012.

To counter competition from NewSpace[27] and in particular SpaceX,[28] Europe decided in late 2014 to develop Ariane 6[29] and Vega-C.[30] The objective is to offer launch services for civil, commercial, and military markets at a better price, but with reliability and constant level of services.[31]

---

[23]Soyuz, *Technical Overview*, April 18, 2019: http://www.arianespace.com/wp-content/uploads/2019/04/ARIANESPACE-ENG-FLYER-SOYUZ-APRIL2019-WEB.pdf (last accessed May 24, 2019).

[24]Russian Space Web, http://www.russianspaceweb.com/kourou_els.html

[25]Room The Space Journal, *France, Europe and Russia—Two Decades of Space Launch Cooperation*, Issue 2(8) 2016: https://room.eu.com/article/france-europe-and-russia-two-decades-of-space-launch-cooperation (last accessed May 24, 2019). See also, SpaceNews, *Arianespace launches ESA's CHEOPS satellite to study exoplanets*, December 18, 2019: https://spacenews.com/arianespace-launches-esas-cheops-satellite-to-study-exoplanets/.

[26]Vega, *Technical Overview*, April 18, 2019: http://www.arianespace.com/wp-content/uploads/2019/04/ARIANESPACE-ENG-FLYER-VEGA-APRIL2019-WEB.pdf (last accessed May 24, 2019); see also AVIO website: http://www.avio.com/en/news-en/avio-contract-signed-with-arianespace-for-10-vega-and-vega-c-launchers/ (last accessed May 24, 2019).

[27]Opinion of Air and Space Academy, *Future of European Launchers*, n° 9, 2019, pp. 6–7: https://academieairespace.com/wp-content/uploads/2019/05/AAE_Opinion9_FutureEuropeanLaunchers Web_EN.pdf (last accessed May 24, 2019).

[28]Ibid., pp. 10–12.

[29]Ariane 6, *Users's Manual*, March 2018: http://www.arianespace.com/wp-content/uploads/2018/04/Mua-6_Issue-1_Revision-0_March-2018.pdf (last accessed May 24, 2019).

[30]Business Insider, *Voici les 3 défis majeurs à relever pour qu'Ariane 6 puisse tenir tête à la concurrence de SpaceX*, February 2019: https://www.businessinsider.fr/fusee-ariane6-defis-pour-rivaliser-spacex-rapport-cour-des-comptes-2019 (last accessed May 24, 2019); see also Vega C, User's Manual, May 2018: http://www.arianespace.com/wp-content/uploads/2018/07/Vega-C-user-manual-Issue-0-Revision-0_20180705.pdf (last accessed May 24, 2019); see also Opinion of Air and Space Academy, *op.cit.*, pp. 8–9.

[31]Opinion of Air and Space Academy, *op.cit.*, pp. 15–17.

3 The Impact of Space Sector on French Guyana Economy

The Guyana Space Center plays a crucial role from an economic and educational point of view. Indeed, it employs 1700 persons[32] and invests about 55 million euros each year in the infrastructures modernization, of which 40–50% are injected into the Guianese economy. Space activity employs 16% of the active population (about 9000 direct and indirect jobs), according to the *Institut national de la statistique et des etudes économiques* (INSEE).[33]

Moreover, since the creation of the Guyana Mission in 2000,[34] CNES has invested 90 million euros over the last 15 years to co-finance 3000 projects and create 3900 long term jobs. CNES also co-finances approximately 26.4 million euros in "state-region plan" contract, and signed various agreements with 17 French Guyana's municipalities. Finally, it contributes to the educational background of students (scholarship, partnership with the University of Guyana)[35] as well as digital development of the territory (telemedicine, participation in the submarine cable project). *"The Spaceport can develop only if Guyana develops"* said former director Bernard Chemoul in February 2016.[36]

In fact, space activity irrigates the Guyanese economy in two forms.[37] As a high-tech business, it generates a high added value. In a sparsely populated Guyanese territory, no other activity generates so much wealth, also because economic benefits have an impact on a variety of subcontractors in various sectors of the economy. For these subcontractors, the presence of the space activity brings stable customers, contracts and a better visibility on their own activity. Finally, all salaries paid for these activities are injected into the Guyanese economy. Employees spend a large part of their income on housing, transport or trade, which benefits companies in the local sector in Guyana.

Hence, the place of the space sector in the Guyanese economy remains very important. It still generates 15% of the added value of the territory,[38] and the main creators of labor are the members of the *"Union des Employeurs de la Base Spatiale"* (UEBS). This runoff from the space sector benefits a multiplicity of

---

[32]La Dépêche, *Centre Spatial Guyanais: 75% des salariés sont des Guyanais*: https://www.ladepeche.fr/article/2017/04/06/2551348-75-des-salaries-sont-des-guyanais.html (last accessed May 24, 2019).

[33]La Croix, *La base de Kourou, locomotive économique de la Guyane*, Mars 2017: https://www.la-croix.com/Economie/France/La-base-Kourou-locomotive-economique-Guyane-2017-03-28-1200835391 (last accessed May 24, 2019).

[34]*'Accompagnement au développement'*: http://www.cnes-csg.fr/web/CNES-CSG-fr/9760-aide-au-developpement.php (last accessed May 24, 2019).

[35]*'Accompagnement au développement'*, *Education et Formation*: http://www.cnes-csg.fr/web/CNES-CSG-fr/11534-education-et-formation.php (last accessed May 24, 2019).

[36]Ibidem.

[37]INSEE Dossier, *L'impact du spatial sur l'économie de la Guyane*, n° 5, 2017: https://www.insee.fr/fr/statistiques/3182000 (last accessed May 24, 2019).

[38]Ibidem.

actors. Certainly, the heart of the space activity is provided by the industrial companies of the space base, whose production activity is sometimes carried out largely in "metropolitan" France or in another ESA Member States. This very important impact is largely due to the production volume of the Space Center itself: direct effect via CNES, Arianespace and the security forces, and indirect effect via UEBS manufacturers.

Space activity should not slow down with the new program Ariane 6 formalized in December 2014.[39] In addition, the rise in investment expenditure and research by 2020 for the realization of new infrastructures will have a favorable impact on the Guyanese economy. Lastly, CNES provides important historic management, ranging from the development and maintenance of the Salvation Islands, to the maintenance of several hotels and a large housing stock in the municipality of Kourou (approximately 250 dwellings).[40]

In addition, as a major space transportation company based in French Guyana, Arianespace makes also a significant contribution to the local economy and community.[41] Arianespace's activities at the GSC launch base generate some 1700 jobs at nearly 40 different firms. In turn, these direct jobs help create many "indirect" jobs in the local economy. Moreover, each Ariane 5, Soyuz or Vega launch brings engineers and managers from other companies to French Guyana to take part in launches and satellite operations, boosting the local economy.

Finally, the Arianespace educational outreach programs mainly target students in secondary and higher education.[42] Each year, this includes the award of three scholarships, so students can continue in higher education. Actions in the cultural sphere center on the annual Carnival, which is one of the biggest celebrations in French Guyana. Arianespace also supports a number of scientific initiatives, including the "*Cayenne à la conquête des étoiles*" event,[43] which invites the public to take part in various activities including workshops and conferences on astronomy.

---

[39]SpaceNews, *ArianeGroup says Ariane 6 enters crucial development phase as French auditor warns against SpaceX*, March 8, 2019: https://spacenews.com/arianegroup-says-ariane-6-enters-crucial-development-phase-as-french-auditor-warns-against-spacex/ (last accessed May 24, 2019).

[40]'*Accompagnement au développement*', *Aménagement du territoire*: http://www.cnes-csg.fr/web/CNES-CSG-fr/11533-amenagement-du-territoire.php (last accessed May 24, 2019).

[41]Arianespace, *Corporate Social Responsibility Report* (2014–2015), p. 15: http://www.arianespace.com/wp-content/uploads/2016/03/CSR_report_2014_2015_FR.pdf (last accessed May 24, 2019).

[42]Ibidem.

[43]Cayenne City website: http://www.ville-cayenne.fr/vie-municipale/a-la-conquete-du-ciel-et-des-etoiles/ (last accessed May 24, 2019).

4 Conclusion

Space activities support undoubtedly the economic development of the French Guyana, and space entities such as the CNES[44] and Arianespace are especially engaged in educational programs for the country.[45]

Moreover, it is of utmost importance to take into consideration the disparities inside the country which still exist, for instance the problem of access to drinking water, electricity, internet or mobile network, or the lack of facilities for the circulation of goods and peoples which is always extremely difficult. There is a huge paradox in French Guyana: on the one hand, the Kourou Space Center that is a strong level of attractiveness; on the other hand, infrastructures that do not follow the demographic development of the region.[46]

As highlighted, the presence of space sector generates wealth by contributing to local economic development and education.[47] Indeed, the center represents 9000 direct and indirect jobs and 15% of the country's GDP territory. In addition, from 30 to 40% of tenders benefit to Guyanese companies. The coming of new launchers will have a positive impact on local activities. That's why the Kourou *Institut Universitaire de Technologie* and the Guyana University are developing training courses related to space activities in order to enable young people to benefit from the space industry, which is a promising sector for the future.[48]

Anne-Sophie Martin is a Doctor of Law. She is currently pursuing research in the field of space law and policy at the Political Sciences Department of the Sapienza University of Rome (Italy). She is particularly interested in the legal issues raised by the development of new space activities, and by the adaptation of the law and policy.

---

[44]SpaceNews, French Space Agency Pledges 10 million euro Boost to French Guiana Economy, July 27, 2018: https://spacenews.com/french-space-agency-pledges-10-million-euro-boost-to-french-guiana-economy/ (last accessed May 24, 2019).
[45]CNES MAG, *Guyane, Une Base Spatiale Pas Comme les Autres*, n° 78, Novembre 2018, p. 23.
[46]French Senate, *Séance du 11 décembre 2014*, pp. 10094–10095: http://www.senat.fr/seances/s201412/s20141211/s20141211.pdf (last accessed May 24, 2019).
[47]CNES MAG, *op.cit.*, pp. 22–23.
[48]French Senate, *Séance du 11 décembre 2014, op.cit.*

Closing the Digital Divide: General Trends and Driving Factors Behind Internet Usage in Latin America

Christopher Yoon

Abstract

Internet providing satellite systems are often regarded as an invention that could close the digital divide because they provide Internet access to remote areas. The following chapter analyzes the digital divide and its driving factors by conducting a quantitative empirical analysis of Internet usage in Latin America. For this, the empirical analysis will draw upon theoretical hypotheses formulated in the scholarly literature, which includes independent variables such as economic income, free markets, access to electricity, urbanization, literacy or corruption. To test these hypotheses, the empirical part will apply a regression tree algorithm that allows us to predict unknown values of Internet usage and to identify important variables contributing to Internet access. The study concludes that the digital divide is considerably closing, at least in quantitative terms. Moreover, the empirical evidence confirms the conventional wisdom on the digital divide, arguing that economic strength is a particular important factor that explains access to the Internet.

C. Yoon (✉)
Institute for Higher Military Leadership, National Defence Academy, Vienna, Austria
e-mail: christopher.yoon@hotmail.com

© Springer Nature Switzerland AG 2020
A. Froehlich (ed.), *Space Fostering Latin American Societies*, Southern Space Studies, https://doi.org/10.1007/978-3-030-38912-3_7

1   Introduction

According to a recent Wallstreet Journal article, a variety of companies including Amazon, Facebook and others are planning to invest into a gigantic space project with the aim to establish an Internet providing satellite system.[1] While in their early stages Internet providing satellite systems allowed a speed of only a few hundred kb/s, modern satellite network technologies enable a much faster broadband and therefore could play an increasing role in data communication.[2] One of the most convincing arguments for Internet satellite systems is that they can cover very large areas and provide Internet access in remote areas, where terrestrial network systems are not available. However, Amazon and Facebook are not the first companies planning to invest into such a project. In 1996, Daniel Kohn, who was then marketing manager of Teledesic Corporation, argued that fiber connection was mainly limited to governments and large telephone corporations, which is why Teledesic expected a growing market for Internet providing satellite systems.[3] After the company had invested hundreds of millions of dollars, however, Teledesic gave up its ambitious project because fiber connections spread faster than expected and, therefore, the company assumed that the market for Internet providing satellite systems would disappear over time.[4]

Whereas Internet providing satellite systems are already common for military applications, in the commercial sector they have been regarded as a niche product at worst and as a technology with the potential to overcome the digital divide at best. The purpose of the following study is to analyze the digital divide by focusing on general trends of Internet usage in Latin America and its driving factors. In doing so, the study will consist of two parts: the first part comprises a theoretical analysis, which includes a discussion of the scholarly literature and the creation of hypotheses based on the academic literature. The second part contains the empirical analysis, which is composed of three steps: the first step looks at the surface of Internet usage and puts the general trends in Latin America into perspective. In the second step the study dives deeper into the driving factors behind Internet usage by analyzing scatterplots and correlation pairs. The third step is the main part of the empirical analysis and applies a regression tree algorithm that will help us to predict Internet usage in Latin America and to identify important factors that explain Internet usage. The study concludes, first, that the digital divide is closing at the

---

[1]Wallstreet Journal (2019) *Hate Your Internet Provider? Look to Space*, Christopher Mims, 10 April, https://www.wsj.com/articles/hate-your-Internet-provider-look-to-space-11554897532 [accessed on 1.6.2019].
[2]Botta, Alessio & Pescapé, Antonio (2013) *New generation satellite broadband Internet services: should ADSL and 3G worry?* in: The 5th IEEE International Traffic Monitoring and Analysis Workshop, pp. 399–404.
[3]Kohn, Daniel M. (1996) *Policy Challenges and Opportunities for Global Mobil Personal Communications by Satellite: The Teledesic Viewpoint*, International Telecommunication Union, https://www.itu.int/newsarchive/wtpf96/paper2.html [accessed on 1.6.2019].
[4]Smith, Ernie (2017) *We were promised skynet*, in: Tedium, 30 March, https://tedium.co/2017/03/30/satellite-internet-challenges-history/ [accessed on 1.6.2019].

global and regional level to considerable extent and, second, that the empirical evidence confirms the conventional wisdom in the digital divide literature, arguing that economic strength determines access to the Internet.

The chapter will start with a discussion of the main theories in the scholarly literature, focusing on the concept of digital divide and the factors explaining access to the Internet. As it will turn out, the concept of digital divide does not represent a coherent theory but rather a set of different perspectives and explanations. The section following the discussion of the scholarly literature will build the hypotheses and will outline the theoretical foundations of the empirical analysis. In a next step, the data used in the empirical analysis will be described and specified. The detailed description of the data will be followed by the empirical analysis, which will consist of the three steps described above. Here, the focus of attention will lie on the regression tree algorithm. In the final section, the empirical results will be summed up and discussed.

2   The Current State of the Art

Digitization is considered as one of the mega trends at the global level. This trend, however, has affected different populations and countries at different speeds. The inequality in quantitative and qualitative access to the Internet is commonly described by the concept of digital divide. One of the most prominent definitions of the term "digital divide" was formulated by Jan Van Dijk, who views the digital divide as "the gap between those who do and do not have access to computers and the Internet".[5] While the digital divide is often regarded as north-south divide, digital inequalities can be also detected in more advanced societies between different social groups and populations. In fact, digital inequalities can go hand in hand with other forms of inequality such as inequalities based on economic strength, social status, ethnicity, gender, geography, age or other groups.[6]

Despite its simple definition, the notion of the digital divide has been full of misconceptions. One misconception, as Van Dijk pointed out, is that the digital divide is not limited to technological inequalities and the lack of physical access.[7] Instead, Van Dijk distinguishes between five different types of digital inequalities: (1) the first type is the lack of *material access*, which means that limited access to

---

[5]Van, Dijk J. (2006) *The Network Society: Social Aspects of New Media*, 2nd edition, SAGE, London, pp. 178.

[6]See James, Jeffrey (2003) *Bridging the Global Digital Divide*, Edward Elgar, Cheltenham, Chap. 1. Witte, James C. & Mannon, Susan E. (2010) *The Internet and Social Inequalities*, Routledge, New York & London. For an analysis of the connection between race/ethnicity and the digital divide see Nakamura, Lisa (2002) *Cybertypes: Race, Ethnicity, and Identity on the Internet*, Routledge, New York & London. Or see Campos-Castillo, Celeste (2015) *Revisiting the first-level digital divide in the United States: gender and race/ethnicity patterns, 2007–2012*, in: Social Science Computer Review, Vol. 33, No. 4, pp. 423–439.

[7]Van, Dijk J. (2006) *Digital divide research, achievements and shortcomings*, in: Poetics, Vol. 34, pp. 221–235.

the Internet derives from the lack of computers and networks; (2) the second digital inequality grows out of limited *skill access*, which means that individuals do not have access to the Internet due to the lack of knowledge, skills and competencies that are necessary to use the Internet; (3) a third type of digital inequality is a lack of *mental access*, which describes the absence of an elementary digital experience; (4) another form of digital inequality is the lack of *usage access*, which means that individuals do not have access to the Internet because they lack meaningful and profitable options to use the Internet for their own good[8]; (5) a fifth inequality stressed by Van Dijk is the inequality in *motivational access*, which implies that different individuals have different degrees of motivation to access the Internet due to a general rejection of the digital medium, the lack of time or because of other reasons.[9]

While the digital divide was originally criticized of being too descriptive, newer approaches have increasingly propose normative theories of why the digital divide occurs. As a starting point, Van Dijk argues that traditional philosophies are still relevant in analyzing the digital divide. Theories of the Marxian tradition, as he points out, would argue that the digital divide can be explained in terms of pocession. Max Weber would insist that the digital divide is a matter of profession and its inherent status and Georg Simmel and Ralf Dahrendorf would claim that the digital divide might be explained by different forms of power and personal relationships.[10] Newer studies argue that traditional inequalities cause or reinforce the digital divide because the access to the Internet considerably depends on the availability of economic, political, social and human capital. According to the stratification hypothesis, for instance, the lack of these different types of capital represent a barrier for entering the digital sphere.[11] Similar to the stratification hypothesis, the so-called *complementary hypothesis* claims that privileged groups can further enhance their social position by using social networks for complementary purposes.

In contrast to the pessimistic view of the stratification hypothesis, there are a variety of theories that can be described as convergence-oriented, arguing that countries, which lagged behind in terms of access to the Internet and computers, increasingly catch up with more advanced societies. What most convergence theories have in common is that they emphasize the importance of technology and knowledge sharing in catching up with more advanced societies. According to the diversification hypotheses, for instance, not only privileged groups benefit from digitization, but also underprivileged groups who can overcome traditional inequalities by using digital technologies.[12] The diversification hypothesis assumes

---

[8]Van Dijk, Jan. & Hacker, Ken (2003) *The digital divide as a complex, dynamic phenomenon*, in: The Information Society, Vol. 19, Nr. 4, pp. 315–326.

[9]Van, Dijk J., *Digital divide research, achievements and shortcomings*, pp. 226.

[10]Ibid., pp. 223.

[11]DiMaggio, Paul & Garip, Filiz (2012) *Network effects and social inequality*, in: Annual Review of Sociology, Vol. 38, pp. 93–118.

[12]Litwin, Howard & Shiovitz Ezra, Sharon (2011) *Social network type and subjective well-being in a national sample of older Americans*, in: The Gerontologist, Vol. 51, No. 3, pp. 379–388.

that digital networks can help in compensating weak positions in the digital space by transforming weak ties to other individuals into economic, social and other types of capital. Furthermore, digital networks offer new opportunities in gaining information, for instance, when it comes to accessing job opportunities, using online learning programs or digital tools that can help in managing everyday life, by which marginalized individuals can empower themselves and overcome their underprivileged position.[13] This is not only true for the individual level. The Internet also allows technology and knowledge transfer from developed countries to peripheral countries, which has a positive effect on their economy.[14]

One of the prominent examples of the stratification hypothesis has been Africa. Fuchs and Horak, for instance, argue that "the digital divide is a very pressing problem for Africa" and conclude that "most African countries are excluded from the information society".[15] Their study, which was published in 2008, identified 57 African countries, in which less than 1% of the surveyed population had access to computers and Internet technology. The reason why the theory of digital divide has been the most prominent perspective on Africa is that "Africa is the least-developed region in the world when income, school enrolment and life expectancy are taken into account [...] Thus, Africa is the continent most affected by poverty and underdevelopment. Africa also lags behind the rest of the world in terms of key indicators of the information society, such as subscriptions to the Internet."[16] Newer research, however, indicates that this account is not uncontested. In fact, there is a growing number of researchers in recent years favoring a more balanced view on digitization in Africa. Although the majority of African countries can be still classified as "digital desserts", Africa is a heterogeneous continent with an increasing number of emerging countries that become "digital oases".[17] There is strong evidence for a positive relationship between economic strength and digital access[18] and since a number of African countries have achieved promising growth rates, a closing digital divide is, at least in quantitative terms, not surprising.

Similar trends can be observed in the Latin American region. For instance, a study on smartphone and Internet usage by the Pew Research Center concludes that many countries in the developing world, including Latin American countries are

---

[13]Mesch, Gustavo; Mano, Rita & Tsamir, Judith (2012) *Minority status and health information search: A test of the diversification hypothesis*, in: Social Science and Medicine, Vol. 75, No. 5, pp. 854–858.

[14]Baddeley, Michelle (2006) *Convergence or Divergence? The Impacts of Globalisation on Growth and Inequality in Less Developed Countries*, in: International Review of Applied Economics, Vol. 20, No. 3, July, pp. 391–410.

[15]Fuchs, Christian & Horak, Eva (2008) *Africa and the digital divide*, in: Telematics and Informatics, Vol. 25, pp. 99–116.

[16]Bornman, Elirea (2016) *Information society and digital divide in South Africa: results of longitudinal surveys*, in: Information, Communication & Society, Vol. 19, No. 2, pp. 264–278.

[17]Wentrup, Robert (2016) *Digital oases and digital deserts in Sub-Saharan Africa*, in: Journal of Science & Technology Policy Management, Vol. 7, No. 1, pp. 77–100.

[18]Evans, Olaniyi (2019) *Repositioning for increased digital dividends: Internet usage and economic well-being in Sub-Saharan Africa*, in: Journal of Global Information Technology Management, Vol. 22, No. 1, pp. 47–70.

catching up with advanced societies.[19] Furthermore, empirical studies provide evidence that social media played an important role in political protests and participation.[20] These findings are supported by a comprehensive quantitative analysis by Raul Katz, arguing that Latin American countries undergo positive trends towards digitization.[21] Using a ranking and clustering algorithm, the study finds that Latin American countries might not be classified as advanced countries such as South Korea, Norway, Denmark, Switzerland or the United Kingdom, but they can be clustered around the second and third highest level of digitization in particular transitional countries and emerging countries. In other words, Latin American countries are doing worse than highly industrialized countries but better than developing countries in Africa. In fact, most countries in Latin America can be compared with emerging countries in Asia and the Middle East.

As the literature already indicated, one of the most important factors explaining the digital divide is the economy.[22] The stronger the economy, the better the availability of financial capital for governments, companies and households to invest into infrastructure and digital technologies. At the individual level, studies emphasize socio-economic variables such as general income, employment status or education.[23] In addition to economic explanations, gender studies assume that digital inequalities evolve along gender lines, some of which found evidence that women tend to have a lower frequency of Internet usage, a lower intensity of Internet usage and a narrower range of online skills and activities.[24] In contrast to this view, research also shows that the digital divide between men and women is closing and that age, profession or a certain life-course are better predictors for digital adoptions than gender.[25]

[19]Poushter, Jacob (2016) *Smartphone Ownership and Internet Usage Continues to Climb in Emerging Economies*, Pew Research Center, 22 February, http://s1.pulso.cl/wp-content/uploads/2016/02/2258581.pdf [accessed on 30.6.2019].

[20]Valenzuela, Sebastian; Somma, Nicolas M.; Scherman, Andres & Arriagada, Arturo (2016) *Social media in Latin America: deepening or bridging gaps in protest participation?* in: Online Information Review, Vol. 40, No. 5, pp. 695–711.

[21]Katz, Raul L.; Koutroumpis, Pantelis & Callorda, Fernando (2013) *The Latin American path towards digitization*, in: Info, Vol. 15, No. 3, pp. 6–24.

[22]James, Jeffrey (2003) *Bridging the Global Digital Divide*. Edward Elgar, Cheltenham, see Chap. 1.

[23]Witte, James C. & Mannon, Susan E. (2010) *The Internet and Social Inequalities,* Routledge, New York & London.

[24]Robinson, Laura; Cotten, Shelia R.; Ono, Hiroshi; Quan-Haase, Anabel; Mesch, Gustavo; Chen, Wenhong; Schulzg, Jeremy; Haleh, Timothy M. & Stern, Michael J. (2015) *Digital inequalities and why they matter*, in: Information, Communication & Society, Vol. 18, No. 5, pp. 569–582.

[25]Yu, Rebecca P.; Ellison, Nicole B.; McCammon, Ryan J. & Langa, Kenneth M. (2016) *Mapping the two levels of digital divide: Internet access and social network site adoption among older adults in the USA*, in: Information, Communication & Society, Vol. 19, Nr. 10, pp. 1445–1464.

3 Theoretical Background

The theoretical background is structured into two parts, of which the first one's focus of attention lies on hypothesis building and of which the second one provides the mathematical background of the regression tree algorithm. In the first part, the variables, which were discussed in the previous section or which were widely used in the empirical studies, will be translated into hypotheses. While most of these variables were subject of the scholarly literature, some of them were rather underrepresented. The second part of the theoretical background will describe the model, which will be used to test the hypotheses.

a. Hypothesis building

In spite of the empirical nature of quantitative data science, data patterns desperately require theoretical explanations. These theoretical explanations are not only important to determine the best possible explanation of competing theories at the academic level, but also to solve concrete problems at the practical level. Moreover, the development of hypotheses help researchers to choose which data will be accessed and aggregated.

Apart from the time series analysis, in which the focus of attention will lie on regions, the items of the data in the following study represent countries. While the main dependent variable will be Internet usage measured by Internet users in % of the total population, the independent variables include a variety of factors that were identified in the scholarly literature. The functions between the independent and dependent variables can have different shapes. Some might show a linear relationship, other functions can appear as u-curves, inverted u-curves or as pareto distributed functions. Here, the hypotheses assume linear relationships, although their exact shapes will be investigated in the empirical part in more detail. Moreover, relationships can have parametric or non-parametric distributions, of which the latter means that the errors are not normally distributed and usually increase at the higher end of the spectrum. Another consideration is that causal effects of independent variables on the dependent variable can be bidirectional. The following analysis, however, will largely ignore bidirectional causality since the research interest focuses on variables affecting Internet usage rather than on the impacts of Internet usage on these variables. A problem that is very common in statistical correlation and regression analysis is multicollinearity, which occurs if the independent variables correlate with each other. As a result, regression coefficients cannot be interpreted as the change of the dependent variable Y if the main independent variable X_1 changes by 1 unit and the control variables X_n are held constant. Since the main purpose of the decision tree algorithm is prediction, multicollinearity can be also ignored. Moreover, the decision tree algorithm automatically excludes variables that provide no additional predictive power.

The first hypothesis represents the stratification hypothesis and analyzes the general trends in Internet usage. The hypothesis assumes that Latin American countries are catching up with advanced societies, at least in quantitative terms. Thus, the first hypothesis can be formulated as follows:

- H_1: Countries in the Latin American and Caribbean region show a positive trend in terms of Internet usage and have been able to catch up with more advanced countries in the last few years. The more time passed by, the higher the share of Internet users.

The second hypothesis describes the relationship between Internet usage and a variable that is often regarded as one of the main factors in the theoretical and empirical literature in particular the economy. As indicated by the academic literature, the digital divide coincides, at least to a considerable extent, with the economic performance because digital infrastructure and devices pose relatively expensive technologies especially for developing countries. As this hypothesis assumes, digitization needs financial capital. Building up Internet infrastructure requires huge private and public investments into network technology. Furthermore, access to the Internet requires technologies such as a computer or a smartphone, software and apps, an Internet provider in form of cable, mobile or satellite Internet or electricity—requirements that usually cost money. The performance of the economy can be measured by a combination of different indicators at different levels. At the macroeconomic level, economic performance are reflected in such indicators as the GDP or GNI per capita, GDP growth rate, inflation, unemployment rate, governmental debt or trade balance. At the microeconomic level, indicators of economic performance can include profits, sales, revenues, total factor productivity or business climate indexes. Two indicators will be used in order to test the economic hypothesis: the first indicator will be the general national income (GNI) per capita, which can be interpreted as an indicator at the macroeconomic level as well as at the individual level. The second indicator to test the economic hypothesis will be a business environment index, which represents the strength of the economy at the enterprise level. To put the economic variable into a function, the hypothesis can be read as follows:

- H_2: Digitization in Latin America heavily depends on economic performance. The greater the economic strength of a country, the higher the share of Internet users.

The third hypothesis assumes that the availability of technology and infrastructure most notably electricity are important factors determining the share of Internet users—a hypothesis that is borrowed from economics. The literature in economic science often refers to this linkage as the energy-growth nexus, assuming that higher electricity outputs go hand in hand with economic growth. Based on a historical analysis of the American economy there is evidence that electricity and mechanization were important variables increasing the total factor productivity in the early 20th century.[26] With a growing energy production and supply in the first

---

[26]Oshima, Harry T. (1984) *The growth of U.S. factor productivity: The significance of new technologies in the early decades of the twentieth century*, in: The Journal of Economic History, Vol. 44, No. 1, March, pp. 161–170.

half of the 20th century, electricity became cheaper. As a consequence, electrical machines were increasingly used to substitute steam-based technologies or hand labor in agriculture and industry and unskilled labor was gradually replaced by mechanized processes and more accurate production processes, both of which resulted in higher total factor productivity per input unit. Due to the access to electricity, industrial production became faster and more efficient not only in the agricultural sector but also in sectors such as mining, transportation or manufacturing.[27] The energy-growth nexus has been a widely popular research area in economics and could be observed in a vast number of emerging countries and markets, too such as China, South Korea, Taiwan, Turkey or other emerging countries.[28] There is strong evidence though that electricity consumption and economic growth can decouple, for instance, if investments are made into energy efficiency.[29] Access to electricity is an important factor contributing not only to the economic performance more generally, but also to digitization more specifically because digital technology heavily depends on the availability of energy and electrical infrastructure. Therefore, the third hypothesis focuses on electricity and can be read as follows:

- H_3: The access to the Internet highly depends on the access to electricity. The higher the share of those having access to electricity, the higher the share of Internet users.

As already mentioned in the previous section, another factor often regarded as one of the main prerequisites for Internet usage is literacy and education. Access to the Internet does not only require traditional skills such as writing and reading, but also newer types of skills necessary to use digital technologies such as technical skills, programming skills or knowledge on how to conduct business on the Internet.[30] Furthermore, Internet content is often not available in local languages, which is why Internet usage usually requires substantial English proficiency. The following hypothesis will test the variable literacy:

---

[27]Ibid., pp. 161–170.

[28]For a general discussion of the energy-growth nexus see Ozturk, Ilhan (2010) *A literature survey on energy-growth nexus*, in: Energy Policy, Vol. 38, pp. 340–349. Specific case studies can be found in Seung-Hoon & Jung, Kun-OhJung (2005) *Nuclear energy consumption and economic growth in Korea*, in: Progress in Nuclear Energy, Vol. 46, No. 2, pp. 101–109. See also Oha, Wankeun & Lee, Kihoon (2004) *Energy consumption and economic growth in Korea: testing the causality relation*, in: Journal of Policy Modelling, Vol. 26, No. 8–9, pp. 973–981. For China see Cui, Huanying (2016) *China's Economic Growth and Energy Consumption*, in: International Journal of Energy Economics and Policy, Vol. 6, No. 2, pp. 349–355. For an analysis of the energy-growth nexus in Africa see Odhiambo, Nicholas M. (2009) *Electricity consumption and economic growth in South Africa: A trivariate causality test*, in: Energy Economics, Vol. 31, pp. 635–640.

[29]Shiu, Alice & Lam, Pun-Lee (2004) *Electricity consumption and economic growth in China*, in: Energy Policy, Vol. 32, pp. 47–54.

[30]Van Dijk, Jan. & Hacker, Ken, *The digital divide as a complex, dynamic phenomenon*, pp. 315–326.

- H_4: The usage of digital technologies considerably depends on literacy and education. The higher the literacy rate of a country, the higher the share of Internet users.

The fifth hypothesis will explore the relationship between Internet usage and urbanization. Similar to the access to electricity, urbanization has been often mentioned as a factor within the context of economic development. In the late 19th century, the economist Alfred Marshall published one of the first comprehensive studies on the linkage between urban areas and economic development.[31] There are a variety of reasons why urbanization might coincide with economic growth. First, cities are concentrated areas with a higher availability of physical infrastructures such as electric power grids, transportation or communication. Furthermore, urban areas provide more managerial resources and a larger labor force than rural areas. Moreover, urban areas might encourage knowledge spill-overs and imply a better access to information and networks.[32] As the rapid urbanization in developing and emerging countries indicate, however, urbanization does not necessarily go hand in hand with economic growth because the population of a city can grow faster than its economy. In the following analysis, the urbanization-growth nexus will be applied to the issue of Internet usage, arguing that higher urbanization rates coincide with higher shares of Internet users. The reasons for this assumption are similar to those of why urbanization can contribute to economic growth: better access to digital technologies, the availability of better and cheaper electrical infrastructure and knowledge spill-overs. As a result, from these considerations, the fifth hypothesis can be formulated as follows:

- H_5: Internet usage requires concentrated areas with access to digital technologies, knowledge spill-overs, infrastructure and other economic factors. The higher the urbanization rate of a country, the higher the share of Internet usage of the total population.

Another important factor the previous section did not mention, but which is often believed to have an impact on economic development, is the performance of governments and institutions and their capability to provide public goods. A study on public management in Africa, for instance, showed that the development of the economy, infrastructure and education system significantly depend on political management capabilities and "good governance" of the ruling elite.[33] Wolde-Rufael, for instance, sees a connection between Africa's energy scarcity and the countries' macroeconomic mismanagement, arguing that "Without improving

---

[31]Marshall, Alfred (1890) Principles of Economics: An introductionary volume, 8th Edition, MacMillan and Co., London.
[32]Henderson, Vernon (2003) *The Urbanization Process and Economic Growth: The So-What Question*, in: Journal of Economic Growth, Vol. 8, pp. 47–71.
[33]Evans, Olaniyi (2018) *Digital government: ICT & public sector management in Africa*, in: New Trends in Management: Regional, Cross-Border and Global Perspectives, Włodzimierz, S. (Ed.), London Scientific Publishing, London.

the management of the economy and reducing the role of the state that has been blamed for Africa's economic ills, it is difficult to envisage how the energy challenges facing African countries can be addressed."[34] Since there is no variable in the data measuring the performance of governments and political institutions directly, the following study will use an indicator associated with functioning political institutions and governmental performance in particular the level of corruption. In order to analyze this variable, the following hypothesis will be tested:

- H_6: Digitization considerably depends on good political management and low corruption. The better the political management of a country, the higher the share of Internet users.

An additional factor, which has gained prominence in the scholarly literature over the last few years, is the linkage between Internet usage and democracy—a relationship that has often been described by two fundamentally different views. Whereas the first view describes the Internet as "liberation technology", which empowers political movements that are anti-government, the second view considers the Internet as a "repression technology", arguing that modern information and communication technologies are just another tool for autocrats to control and surveil their populations.[35] As for the first view, for instance, a study on the impact of social media on the Arab uprising argues that new media "played a critical role especially in light of the absence of an open media and a civil society".[36] Another study on the impact of social media on the Arab uprising argues that the Internet increased the speed of information flows, strengthened ties between political activists and leaders of the opposition and contributed to solidarity between protesters and the rest of the world. Moreover, social media helped in gaining sympathy and in spreading social unrest to other countries.[37]

The second view, by contrast, argues that modern information and communication technologies are "repression technologies", which means that governments tend to control the domestic information environment and narrative in order to consolidate and fortify autocratic power.[38] The following hypothesis will assume that the type of the political system has a positive impact on the spread of network technology. Democracies, as the argument goes, adopt Internet technology much faster than other regime types because of a variety of reasons. As Milner pointed out, democracies provide safety nets for those, who suffer from technological

---

[34]Wolde-Rufael, Yemane (2006) *Electricity consumption and economic growth: a time series experience for 17 African countries*, in: Energy Policy, Vol. 34, pp. 1106–1114.

[35]Rod, Espen Geelmuyden & Weidmann, Nils B. (2015) *Empowering activists or autocrats? The Internet in authoritarian regimes*, in: Journal of Peace Research, Vol. 52, No. 3, pp 338–351.

[36]Khondker, Habibul Haque (2011) *Role of the New Media in the Arab Spring*, in: Globalizations, Vol. 8, No. 5, pp. 675–679.

[37]Eltantawy, Nahed & Wiest, Julie B. (2011) Social Media in the Egyptian Revolution: Reconsidering Resource Mobilization Theory, in: International Journal of Communication, Vol. 5, pp. 1207–1224.

[38]Rod, Espen Geelmuyden et al. *Empowering activists or autocrats? The Internet in authoritarian regimes*, pp. 338.

change. Moreover, democratic governments are more concerned about the state of the economy, which is why they do not detain technological change. Even more importantly, democracies are characterized by more freedoms. Based on these considerations, the second hypothesis goes like this:

- H_7: Digitization and the spread of Internet technology are more likely to evolve within democratic regimes. The more democratic a country's regime, the higher the share of Internet users.

b. **Mathematical background**

Whereas the previous section dealt with the development of the hypotheses, the following subsection will present the mathematical fundamentals of the algorithm, which will test the hypotheses. The algorithm will be a regression tree model, which allows us to perform two main tasks. The first purpose of the algorithm is to predict Internet usage of countries in a testing dataset based on a model that will be developed by a training dataset. The second task of the algorithm is to identify the main independent variables that can explain and predict Internet usage.

Regression tree models are a special type of decision tree algorithms. Decision tree models are machine-learning algorithms, whose standard approach is described in Fig. 1. As can be seen in the diagram, the first step in developing a decision tree algorithm is to split the dataset into a training and a testing dataset. The training dataset is used for the process of induction, in which the model will be trained based on this dataset. What follows is the process of deduction, in which the model will be applied to a test dataset.[39] At the end of the deduction process the performance of the model will be evaluated. If the performance of the model turns out to be accurate, the model could potentially be used for unknown data in real world applications.

Although most decision tree algorithms predict two different classes, for instance "credit default" and "credit non-default", they can be also used to predict more than two classes. As can be seen in Fig. 2, decision trees consist of a flowchart with a root node, internal nodes and leaf nodes. The root node is the starting point of the flowchart and stands for the entire dataset. The internal nodes represent a split of the dataset into two or more homogenous data segments. To split the dataset at an internal node, the algorithm uses a variable that produces the most homogenous data segments after a split. For instance, the C5.0 algorithm uses the concept of entropy as a measure of homogeneity, which, generally speaking, describes the ratio between the number of classes in the data segment. The split of the data segments will be repeated either until the data segments are completely homogenous or until there are no variables left that can produce a homogenous segment. Each node of the decision tree can be also expressed by a rule. For instance, if the

---

[39]Tan, Pang-Ning; Steinbach, Michael & Kumar, Vipin (2006) *Introduction to Data Mining*, Pearson Education, Inc., Boston, Chap. 4.

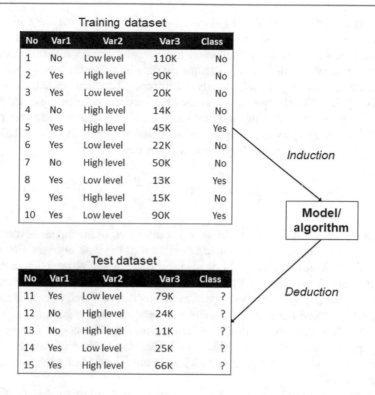

Training dataset

No	Var1	Var2	Var3	Class
1	No	Low level	110K	No
2	Yes	High level	90K	No
3	Yes	Low level	20K	No
4	No	High level	14K	No
5	Yes	High level	45K	Yes
6	Yes	Low level	22K	No
7	No	High level	50K	No
8	Yes	Low level	13K	Yes
9	Yes	High level	15K	No
10	Yes	Low level	90K	Yes

Induction

Model/ algorithm

Test dataset

No	Var1	Var2	Var3	Class
11	Yes	Low level	79K	?
12	No	High level	24K	?
13	No	High level	11K	?
14	Yes	Low level	25K	?
15	Yes	High level	66K	?

Deduction

Fig. 1 This figure illustrates the general approach of classification models and machine learning algorithms more generally

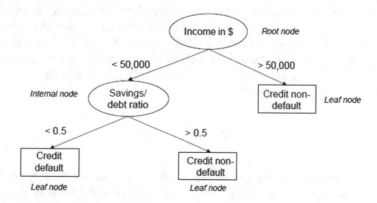

Fig. 2 This graphic shows the basic idea behind a decision tree algorithm classifying credit borrowers into the classes "credit default" or "credit non-default" based on the two independent variables "income" and "savings/debt ratio"

independent variable is higher than x, the dependent variable will have the attribute y. In a next step, these flowcharts with its nodes and leafs are used to predict unknown classes of a testing dataset.

Regression trees are similar to decision trees in so far that they produce a flowchart based on the so-called divide-and-conquer strategy. In contrast to decision tree algorithms, however, regression trees do not predict different classes or factor variables but are used to predict numeric data by using the average value of a variable represented by a leaf. Another difference between decision trees and regression trees is that the splits are not based on the concept of entropy, which only applies to non-numeric dependend variables. Instead, the split in the regression tree in the following analysis draws upon the so-called Standard Deviation Reduction (SDR), which can be described by the following formula:

$$SDR(A, S) = SD(S) - \sum^n \frac{|S_i|}{|S|} SD(S_i)$$

in which the term $SD(S)$ refers to the standard deviation of the values in the overall dataset S, $SD(S_i)$ represents the standard deviation of the data segment after a split on a certain variable and $|T|$ and $|T_i|$ stands for the number of observation in the data segment T and T_i respectively. In other words, this formula calculates the standard deviation of the pre-split standard deviation minus the weighted post-split standard deviation. The more the standard deviation is reduced after a split, the higher the priority the algorithm assigns to the respective variable. The nodes in a regression tree represent rules, too. The only difference is that the outcome of the rule does not determine a class such as "credit default" or "non-default" but rather averages of numeric values.[40]

As soon as the model predicted Internet usage in the testing dataset, the performance of the regression tree model will be evaluated. For this, the evaluation includes four procedures: the first procedure is to compare the summary statistics of the predicted and actual values. The second procedure is to calculate the correlation between the actual values and the predicted values. The higher the correlation coefficient, the better the performance of the model. In the third procedure, the mean absolute error (MAE) will be calculated, which represents the average difference between the actual value and the predicted value, and which can be calculated by the following formula:

$$MAE = \frac{1}{n} \sum_{i=1}^n |e_i|$$

in which n stands for the number of predictions and e_i for the error of prediction i. Note that the vertical strokes in the term $|e_i|$ are not borrowed from set theory and, therefore, do not represent the number of observations, as it is the case in the SDR formula. Rather, they symbolize absolute values, which means that negative values are transferred into positive values. In the fourth procedure, the actual and predicted values will be directly compared and analyzed.

---

[40]Lantz, Brett (2015) *Machine Learning in R: Expert Techniques for Predictive Modeling to Solve All Your Data Analysis Problems*, 2nd Edition, Packt Publishing, Birmingham, UK, Chap. 6.

As compared to usual regression models, regression tree models have the advantage that they better fit to non-linear data. Moreover, linear regression models, in contrast to regression trees, are based on assumptions about the distribution of the data, which often hold not true for data in the real world. The advantage of regression tree models over linear regressions is that they are less dependent on these assumptions. Furthermore, regression trees can handle very complex data with large datasets and different types of variables. Another advantage compared to linear regression models, in which the independent variables need to be predefined, is that regression trees automatically select the independent variables depending on the best possible outcome.

4 Data

In this study, the empirical analysis uses two different datasets. Whereas the first dataset was obtained from the Worldbank, the second dataset is the *Inclusive Internet Index 2019*, which was commissioned by Facebook and generated by the Economist Intelligence Unit.[41] The Worldbank dataset includes time series data combined with cross-sectional data on more than 1,500 economic and social indicators—a dataset that allows us to analyze the temporal dimension of Internet usage. The time series data ranges from 1960 until today and provides data for all countries and regions. Because the dataset treats the regions as single entities, the data will be weighted based on the population sizes of each of the countries.

By contrast, the Inclusive Internet Index provides cross-sectional data focusing on digital and economic indicators and covers the largest 100 countries in terms of the size of their economy. Among these 100 countries, there are 14 countries from the Latin American and Caribbean region, including Argentina, Brazil, Chile, Colombia, Costa Rica, Dominican Republic, Ecuador, El Salvador, Guatemala, Jamaica, Panama, Peru, Uruguay and Venezuela. In contrast to the Worldbank dataset, the analysis of the Inclusive Internet Index is based on country-level data, which is why the data will be not weighted according to the countries' population size. Whereas the time series trends use the Worldbank dataset, the Inclusive Internet Index will help us to explore the driving factors behind Internet usage in more debth. As described in the previous section, the regression tree requires the split of the dataset into a training dataset and a test dataset. While the training dataset will include countries from all regions other than the Latin American and Caribbean region, the test dataset will consist of Latin American and Caribbean countries.

As it is with most data, both datasets have strengths and weaknesses. Although the Worldbank dataset is highly comprehensive, it does not cover all indicators for all countries and years. For this reason the absolute numbers in the dataset do not

---

[41]The dataset of the Worldbank can be downloaded from https://data.worldbank.org/country. The Inclusive Internet Index 2019 is accessible on https://theinclusiveInternet.eiu.com/.

fully reflect the scope of the development, which is why it is not so much about the absolute figures of Internet users than about general trends that can be expressed in relative terms. By contrast, the Economist dataset might be less comprehensive than the Worldbank dataset concerning the number of variables, countries and timespan, however, the advantage of the Inclusive Internet Index over the Worldbank dataset is that there is no missing data. In spite of the quantitative nature of the dataset, it also needs to be considered that the collection of data and the development of the dataset depend on definitions and codes. For instance, the variable Internet usage as a share of a country's population can vary from dataset to dataset because the data can have different underlying definitions. Whereas the Worldbank dataset and the Inclusive Internet Index define an Internet user as someone, who used the Internet at least three months prior to the survey, other sources define a threshold of one year. As a consequence, in some statistics Internet usage appears to be higher than in the more conservative data of the Worldbank and Inclusive Internet Index.

Table 1 lists the main variables of the Inclusive Internet Index, which are part of the empirical analysis. The table also provides information about the type of the variable and how it is defined in the dataset and its codebook.

5 Empirical Analysis

In order to test the hypotheses formulated above, the empirical analysis consists of three parts. The first part will look at the surface of Internet usage and put the digital divide into global perspective. In doing so, the aggregated trends in Latin America will be compared to the developments in other parts of the world. Additionally, a ranking of one hundred countries according to their share of Internet users will be generated and inspected, which allows us not only to compare Latin America with other regions but also specific countries. The second empirical part will take a closer look at the driving factors behind Internet usage and will analyze the relationships between the independent and dependent variables, first, by conducting a correlation analysis and, second, by exploring scatterplots. The third part of the empirical analysis includes the regression tree algorithm, which will help us to predict Internet usage of Latin American countries and to identify to main variables determining the share of Internet users.

a. **Trend and ranking analysis**

Figure 3 shows the general trends of the share of Internet users in percentage of the total populations at the regional level. As can be seen in the figures, network technology has spread rapidly in all regions, however, in some regions with more speed than in others. The leading region in terms of Internet usage is North America with the United States being the main driving force behind it. This is not surprising since most of the research and investment contributing to the development and spread of computers and network technology has been conducted in the United

Table 1 Main variables of the economist's inclusive internet index 2019 dataset used in this study

Variable	Type of variable	Description
Country	String	The country variable displays and identifies the country's name
Internet usage	Numeric	The variable Internet usage is measured by Internet users in \% of population and was obtained from the International Telecommunications Union. Internet users are defined as individuals who used the Internet in the last three months. The Worldbank dataset uses the same definition
GNI per capita	Numeric	Gross national income is measured in current US\$ per person and is calculated by GDP plus net capital flows divided by population size. The data was originally obtained from the Worldbank dataset
Business environment index (BEI)	Interval	The BEI was created by the economist intelligence unit in order to measure the attractiveness of a country for doing business. It includes indicators such as market opportunities, taxes, access to capital or policies towards free markets, enterprise and competition. The Index is measured by a scale of 1–10
Electricity access	Numeric	Electricity access is measured in \% of population and was provided by national household surveys
Literacy rate	Numeric	The level of literacy is measured in \% of population and was obtained from the UNESCO
Urbanization rate	Numeric	The urbanization rate measures the share of the population living in urban areas and was provided by the Worldbank
Corruption index	Interval	The source of the corruption index is Transparency International and is measured by a scale of 1–100, in which the value 1 represents the highest level of corruption
Democracy index	Interval	The democracy index measures the level of democracy of a country, The scale of the variable ranges between 0 and 10, of which 10 means highest level of democracy

States.[42] While Internet usage accelerated rapidly in North America between the mid-1990s and the mid-2000s, growing from 4.6% of the population in 1994 to 74.8% in 2007, the curve flattened after that period but continued to grow with a lower speed to 77% in 2017.

---

[42]Mowery, David C. & Simcoe, Timothy (2002) *Is the Internet a US invention?—an economic and technological history of computer networking*, in: Research Policy, Nr. 31, S. 1369–1377. See also Van den Ende, Jan & Kemp, René (1999) *Technological transformations in history: how the computer regime grew out of existing computing regimes*, in: Research Policy, Nr. 28, S. 833–851.

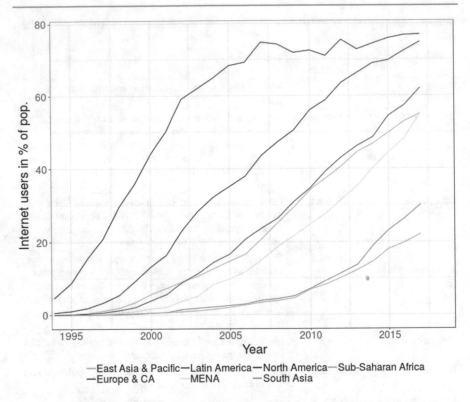

Fig. 3 The graph illustrates the long-term trends of the share of Internet users measured in \% of the total populations at the regional level. The figures are weighted averages based on the countries' population sizes (*Data:* Worldbank)

The rapid spread of Internet usage until 2007 increased the digital divide between the North American region and the rest of the world. However, the adoption of computers and network technology in other regions have increasingly reduced the digital divide since the late 1990s and with accelerating speed since the mid-2000s. While Europe was the second fastest region adopting network technology with a growing share of Internet users from 5.4% in 1998 to 74.9% in 2017, other regions could catch up, too. As a result, the digital divide between North America and Europe, on the one hand, and the rest of the world, on the other, has been gradually reduced. The Latin American region seems to follow a path that is similar to the one of the Asian Pacific region and the Middle East especially since the early 2000s. Today, the Latin American region and the Caribbean countries have the third highest share of Internet users with 62.1% in 2017, while both the Asian Pacific and the Middle Eastern region have a share of Internet users of 55%.

Also, the South Asian and Sub-Saharan regions catch up with more advanced societies in terms of Internet usage, although the Internet started to spread much later and continued to spread with a much slower pace than in other regions. The share of Internet users increased from 5.1% in 2009 to 30.2% in 2017 in South Asia and from 4.5% in 2009 to 22.1% in 2017 in Sub-Saharan Africa. Note, however, that the regions outside North America and Europe were faced with rapid population growth. According to Worldbank data, the population of South Asia grew by 200 Mio. between 2009 and 2017, the Sub-Saharan African population by 205 Mio., the population of the Asian Pacific region by 110 Mio. and even the Latin American and the Middle Eastern region grew by 52 Mio. and 64 Mio. respectively in the same time period. Considering this trend in population size, the figures of the share of Internet users and the closing digital divide between these regions and the more advanced societies is even more astonishing. The growing share of Internet users in spite of rapid population growth means that the velocity of digitization is faster than human reproduction.

Table 2 ranks the largest 100 countries according to their share of Internet users in %, which seems to confirm the Worldbank data in Fig. 3. In the upper third of the ranking most countries are from Europe, North America and a couple of outliers from the Gulf region and the East Asian and Pacific region such as South Korea, Japan, Singapore, Malaysia, Taiwan and Australia. With the exception of Chile, most Latin American and Caribbean countries are listed in the second third of the ranking together with the vast majority of East Asian and Middle Eastern countries. In fact, 12 out of 14 Latin American and Caribbean countries are listed between 35th and 67th rank. African countries and South Asian nations are ranked in the lower third.

Considering the time series analysis and the inspection of the ranking, there is enough evidence that H_1 can be confirmed. In fact, the digital divide between the advanced countries, on the one hand, and Latin American countries, on the other, is closing rapidly since most countries are catching up with Internet usage despite rapid population growths. The same is true for other regions such as the Asian-Pacific region, MENA or even South Asia and Sub-Saharan Africa. What the trend analysis and ranking table also suggest is that Internet usage has a regional (geographic) dimension with countries being clustered around other countries from the same region. There are a variety of explanations for this, most of which could be borrowed from economics such as regional trade institutions, cheaper access to imported goods and products and export markets or stable political environments to transport goods and to conduct trade.

b. **Correlation and scatterplot analysis**

In order to identify and explore the driving factors behind Internet usage, the following part uses correlation and scatterplot analysis. The correlations in Table 3 show that all independent variables have a high or a modest positive correlation with Internet usage and all of which are highly significant. With a correlation coefficient of 0.94, the logarithmized GNI per capita has the strongest relationship with Internet usage. The other variables have high or moderate correlations, too, most of which ranging between 0.75 and 0.82. In fact, the correlation table reveals the importance of urban areas and free markets in gaining access to the Internet.

Table 2 List of the 100 largest economies ranked by the share of Internet users of the total population (Data: Inclusive Internet Index 2017).

Rank	Country	Region	Internet users in %
1	Kuwait	MENA	97.9
2	Denmark	Europe/CA	97.0
3	Sweden	Europe/CA	96.4
4	South Korea	Asia/Pacific	95.0
5	UAE	MENA	94.8
6	United Kingdom	Europe/CA	94.7
7	Qatar	MENA	94.2
8	Switzerland	Europe/CA	93.7
9	Netherlands	Europe/CA	93.1
10	Canada	North America	91.1
11	Japan	Asia/Pacific	90.8
12	Estonia	Europe/CA	88.1
13	Austria	Europe/CA	87.9
14	Belgium	Europe/CA	87.6
15	Finland	Europe/CA	87.4
16	Australia	Asia/Pacific	86.5
17	Spain	Europe/CA	84.6
18	Ireland	Europe/CA	84.5
19	Singapore	Asia/Pacific	84.4
20	Germany	Europe/CA	84.3
21	*Chile*	*Latin America/Car.*	*82.3*
22	Israel	MENA	81.5
23	France	Europe/CA	80.5
24	Malaysia	Asia/Pacific	80.1
25	Saudi Arabia	MENA	80.0
26	Taiwan	Asia/Pacific	79.7
27	Czech Republic	Europe/CA	78.7
28	Oman	MENA	76.8
29	Hungary	Europe/CA	76.7
30	Kazakhstan	Europe/CA	76.4
31	United States	North America	76.1
32	Russia	Europe/CA	76.0
33	Poland	Europe/CA	75.9
34	Portugal	Europe/CA	73.7
35	*Costa Rica*	*Latin America/Car.*	*71.5*
36	*Argentina*	*Latin America/Car.*	*70.9*
37	Greece	Europe/CA	69.0
38	*Uruguay*	*Latin America/Car.*	*66.4*
39	Turkey	Europe/CA	64.6

(continued)

Table 2 (continued)

Rank	Country	Region	Internet users in %
40	*Dominican Republic*	*Latin America/Car.*	*63.8*
41	*Mexico*	*Latin America/Car.*	*63.8*
42	Romania	Europe/CA	63.7
43	Bulgaria	Europe/CA	63.4
44	Jordan	MENA	62.3
45	*Colombia*	*Latin America/Car.*	*62.2*
46	Morocco	MENA	61.7
47	Italy	Europe/CA	61.3
48	*Brazil*	*Latin America/Car.*	*60.8*
49	Iran	MENA	60.4
50	*Venezuela*	*Latin America/Car.*	*60.0*
51	*Ecuador*	*Latin America/Car.*	*57.2*
52	Tunisia	MENA	55.5
53	Philippines	Asia/Pacific	55.5
54	China	Asia/Pacific	54.3
55	South Africa	Sub-Saharan Africa	54.0
56	*Panama*	*Latin America/Car.*	*54.0*
57	Ukraine	Europe/CA	53.0
58	Thailand	Asia/Pacific	52.8
59	*Peru*	*Latin America/Car.*	*48.7*
60	Vietnam	Asia/Pacific	46.5
61	Egypt	MENA	44.9
62	*Jamaica*	*Latin America/Car.*	*44.3*
63	Cote d'Ivoire	Sub-Saharan Africa	43.8
64	Algeria	MENA	42.9
65	Botswana	Sub-Saharan Africa	39.3
66	Ghana	Sub-Saharan Africa	34.6
67	*Guatemala*	*Latin America/Car.*	*34.5*
68	Cambodia	Asia/Pacific	33.9
69	Indonesia	Asia/Pacific	32.2
70	Sri Lanka	South Asia	32.0
71	Namibia	Sub-Saharan Africa	31.0
72	India	South Asia	29.5
73	*El Salvador*	*Latin America/Car.*	*28.9*
74	Sudan	Sub-Saharan Africa	28.0
75	Nigeria	Sub-Saharan Africa	25.6
76	Senegal	Sub-Saharan Africa	25.6
77	Zambia	Sub-Saharan Africa	25.5
78	Myanmar	Asia/Pacific	25.0
79	Cameroon	Sub-Saharan Africa	23.2

(continued)

Table 2 (continued)

Rank	Country	Region	Internet users in %
80	Mongolia	Asia/Pacific	22.2
81	Uganda	Sub-Saharan Africa	21.8
82	Rwanda	Sub-Saharan Africa	20.0
83	Nepal	South Asia	19.6
84	Bangladesh	South Asia	18.2
85	Mozambique	Sub-Saharan Africa	17.5
86	Kenya	Sub-Saharan Africa	16.6
87	Pakistan	South Asia	15.5
88	Ethiopia	Sub-Saharan Africa	15.3
89	Burkina Faso	Sub-Saharan Africa	13.9
90	Tanzania	Sub-Saharan Africa	13.0
91	Angola	Sub-Saharan Africa	13.0
92	Benin	Sub-Saharan Africa	11.9
93	Sierra Leone	Sub-Saharan Africa	11.7
94	Malawi	Sub-Saharan Africa	11.4
95	Mali	Sub-Saharan Africa	11.1
96	Niger	Sub-Saharan Africa	10.2
97	Madagascar	Sub-Saharan Africa	9.8
98	Guinea	Sub-Saharan Africa	9.8
99	Liberia	Sub-Saharan Africa	7.3
100	Congo (DRC)	Sub-Saharan Africa	6.2

The regions assigned to the countries were taken from the Worldbank classification. The rows with Latin American countries are in italics (*Data* Inclusive Internet Index 2019)

Furthermore, literacy and the level of corruption correlate with Internet usage, too, although to a lesser degree than the general income, urbanization and free markets. According to the correlation table, the independent variable level of democracy has the lowest correlation with Internet usage, although a coefficient of 0.51 indicates that the strength of the relationship is small to moderate. Looking at the relationships between the independent variables themselves, the correlation table exhibit that they correlate with each other, which might give us some understanding of how economic development evolves. In fact, there is evidence that the general national income depends on a combination of free markets, access to energy, urbanization, literacy, good governance and to a lesser degree democratic political institutions.

High correlation coefficients do not necessarily mean that the relationships between the variables are linear. In fact, correlations can have different shapes, which is why they need to be combined with a scatterplot analysis.[43] However, the scatterplots in Fig. 4a–f largely confirm the results of the correlation analysis,

---

[43]Vanhove, Jan (2016) *What data patterns can lie behind a correlation coefficient?* 21 November, https://janhove.github.io/teaching/2016/11/21/what-correlations-look-like [accessed on 6.7.2019].

Table 3 Pearson correlation matrix with ordinary correlation coefficients and significance levels (Data: Inclusive Internet Index 2017).

Variables	Internet users	Literacy rate	Urbanization	LogGNI per capita	BEI	Electricity access	Corruption index	Democracy index
Internet users	1.00	0.78***	0.82***	0.94***	0.81***	0.79***	0.75***	0.51***
Literacy rate	0.78***	1.00	0.69***	0.80***	0.61***	0.81***	0.52***	0.56***
Urbanization	0.82***	0.69***	1.00	0.83***	0.59***	0.69***	0.58***	0.70***
Log GNI per capita	0.94***	0.80***	0.83***	1.00	0.80***	0.79***	0.78***	0.42***
BEI	0.81***	0.61***	0.59***	0.80***	1.00	0.59***	0.87***	0.48***
Electricity access	0.79***	0.81***	0.69***	0.79***	0.59***	1.00	0.47***	0.44***
Corruption index	0.75***	0.52***	0.58***	0.78***	0.87***	0.47***	1.00***	0.70***
Democracy index	0.51***	0.56***	0.70***	0.42***	0.48***	0.44***	0.70***	1.00

*** <0.001; ** <0.01; * <0.1

indicating strong relationships between the independent variables and the dependent variable Internet usage. The logarithmized GNI per capita, the business environment index and the urbanization rate are characterized by a very strong linear and parametric relationship with the share of Internet users. In the case of electricity access the graph also shows a linear relationship. Looking at the residuals of the graph, however, the relationship involves a heteroscedastic distribution of the errors, which means in this case that the errors are growing at the higher end of the spectrum. The other independent variables corruption perception index and literacy rate also show some relationship with the dependent variable Internet usage, however, they do not seem to be as linear as the linkages described before. Instead, their relationships look more like parabola. These results are evidence that the

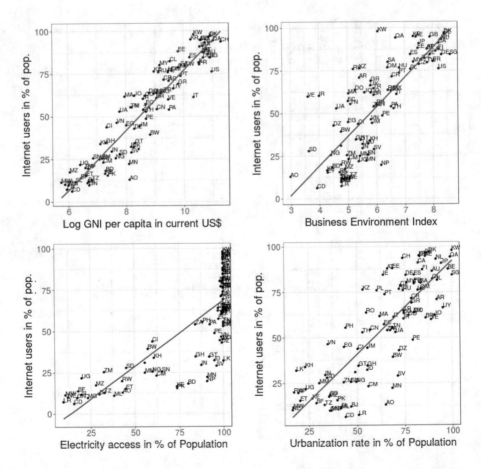

Fig. 4 a–f The following scatterplots illustrate the relationships between the dependent variable share of Internet users and the independent variables GNI per capita, access to electricity, urbanization, literacy, corruption and level of democracy. The data points in the graphs are labelled by the country codes (Data: Inclusive Internet Index 2017)

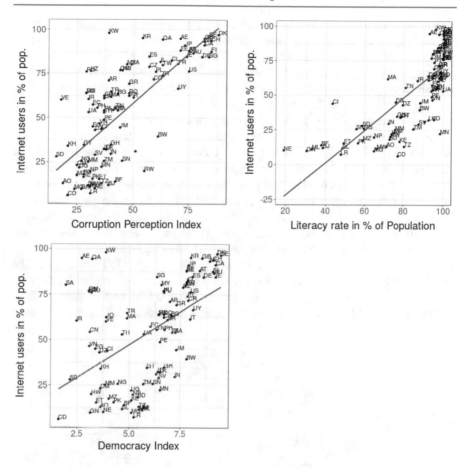

Fig. 4 (continued)

hypotheses H_1–H_6 arguably hold true. In fact, higher general income, a business friendly climate, a better access to electricity, a more urbanized environment, better education and better political management result in a higher share of Internet users of a country's total population. What the scatterplots and the correlation analysis reveal is that the variable GNI per capita is a highly important factor in determining Internet usage. The question here is which variables explain economic strength. As the theoretical part and the correlation analysis suggest, economic strength depends on variables such as business environment and free enterprise, urbanization, infrastructure, literacy or the political management.

c. **Regression tree algorithm**

As already described in the theoretical part, the following section will apply a regression tree algorithm that has two main purposes: predicting the average share

of Internet users of Latin American countries in the testing dataset and identifying
the main variables determining and explaining the share of Internet users. Before
the algorithm can fulfil these tasks, however, the dataset will be split up into a
training dataset and a test dataset. The regression tree algorithm will be trained by
all non-Latin American and non-Caribbean countries that are listed in Table 2. In
contrast to this, the test dataset includes the Latin American and Caribbean countries, whose values will be predicted by the trained model.

Figure 5 presents the regression tree model trained by data of the Non-Latin
American countries in the Inclusive Internet Index. Although all of the independent
variables were included into the training process of the algorithm, the decision tree
only uses two variables to predict the average share of Internet users, in particular
GNI per capita and access to electricity. The rest of the independent variables were
dropped by the algorithm because they did not provide any additional predictive
information. The tree includes a top node, which represents the entire training
dataset and reveals an average Internet usage of 53.31%. Below the top node the
data is split into two data segments by using the variable GNI per capita. The
variable GNI per capita creates two internal nodes, one of which includes the data
segment with a higher or equal GNI per capita of 7,290 US$ and an average share
of Internet users of 81.71% and the other one with a lower GNI per capita than

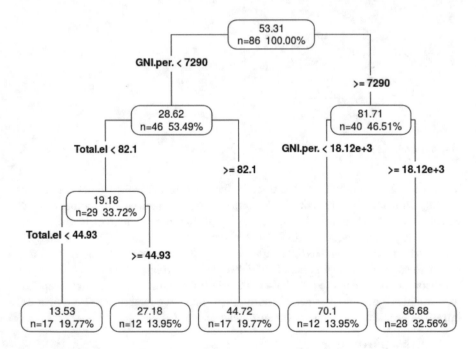

Fig. 5 The regression tree model is based on the training dataset, which will then be applied to
the testing dataset (Data: Non-Latin American countries in the Inclusive Internet Index 2017).

7,290 US$ and an average share of Internet users of 28.62%. The internal node with a higher GNI per capita than 7,290 US$ is split again by the variable GNI per capita, resulting into two leaf nodes, one of which has a higher GNI per capita than 18,120 US$ and an average share of Internet users of 86.68% and one of which lies between 7290 and 18,120 US$ and has on average share of Internet users of 70.1% of the total population. The internal node with a lower GNI per capita of 7,290 US$ is split up into one leaf node and another internal node by the second most important variable in particular access to electricity. In fact, those countries with a GNI per capita lower than 7,290 US$ were subdivided into countries, whose access to electricity is lower than 44.93% with an average Internet usage of 13.53%, whose access to electricity lies between 44.93% and 82.1% with an average Internet usage of 27.18% and those countries with an access to electricity higher than or equal to 82.1% and an average Internet usage of 44.72%.

The regression tree model above will now be applied to the testing dataset in order to predict the share of Internet users in Latin American countries. There are four procedures to evaluate the performance of the algorithm in predicting the dependent variable in the testing dataset. The first procedure is to compare the predicted and actual values shown in Table 4. As can be seen in the summary statistics, the algorithm had its problem to predict the very low and the very high cases. Whereas the algorithm predicted a minimum share of Internet users of 44.72% in the testing dataset, the actual lowest value was 29%. The same is true for the maximum, predicting a maximum value of 70.10% rather than the actual maximum of 82.33%. Overall, this outcome is not that bad because there are only two countries with an actual share of Internet users lower than 44.72%, in particular El Salvador with 28.9% and Guatemala with 34.5%, and one country higher than 70.10%, which is Chile with 82.3%.

The second procedure to evaluate the performance of the model is to calculate the correlation between the predicted and actual values of the testing dataset, which resulted in a correlation coefficient of 0.64. A correlation coefficient of 0.64 means that the regression tree model performed quite well in predicting Internet usage in Latin America and the Caribbean. The third procedure of the model performance evaluation is to calculate the mean absolute error, of which the formula was presented in the theoretical part. The MAE of the predicted values in the testing dataset is 9.51, which means that the average error of the predicted values is 9.51 percentage points. Given the scale between 0 and 100%, an average error rate of 9.51 percentage points seems fairly okay, although there is a high probability that other algorithms such as artificial neural networks or support vector machines would show a better performance than the regression tree algorithm. In fact, the regression tree model shows a better performance with a MAE of 9.51 compared to a MAE of 11.08 if the

Table 4 Summary statistics of the predicted and actual values in the test dataset

	Min.	1st Qu.	Median	Mean	3rd Qu.	Max
Predicted	44.72	44.72	57.41	57.41	70.10	70.10
Actual	29.00	50.05	60.44	57.58	65.77	82.33

Table 5 Comparison of the actual and predicted values of Internet usage in Latin American countries

Country	Actual value	Predicted value	Difference
Jamaica	44.36	44.72	0.36
Argentina	70.96	70.10	0.86
Costa Rica	71.58	70.10	1.48
Uruguay	66.40	70.10	3.70
Peru	48.72	44.72	4.00
Brazil	60.87	70.10	9.23
Venezuela	60.00	70.10	10.10
Guatemala	34.50	44.72	10.22
Chile	82.32	70.10	12.22
Ecuador	57.27	44.72	12.55
Panama	54.00	70.10	16.10
Colombia	62.25	44.72	17.53
Dominican Republic	63.87	44.72	19.15

mean of the actual values is used to predict Internet usage. The fourth procedure to evaluate the performance of the model is to compare the actual and predictive values of the testing dataset directly. As can be seen in the last column of Table 5, there are six countries with a difference between the actual and predicted value lower than the MAE of 9.51 percentage points, while eight countries have a MAE higher than 9.51 percentage points. None of the countries has an error rate higher than 20 percentage points. Nevertheless, there is room for improvement. One of the reasons why the regression tree could have a better performance is that access to electricity, which is one of the two variables predicting Internet usage at the global level, is a somewhat less than perfect variable to predict Internet usage in Latin America because almost all countries in Latin America have an access to electricity higher than 90% of the population and yet deviating shares of Internet users. Thus, if the regression tree model were trained by Latin American countries alone, the model would probably have chosen a variable other than the access to electricity.

As the regression tree analysis indicates, the most important variables to predict the share of Internet users at the global level are the income per capita and access to electricity. Applying the model to the Latin American region, however, access to electricity turned out to be a somewhat less than perfect variable to predict Internet usage because Latin American countries do not vary in terms of electricity access but yet have different degrees of Internet usage. If the decision tree were trained by Latin American countries alone, the algorithm would probably have chosen a variable other than access to electricity to predict Internet usage. The leaf nodes of the generated regression tree are almost identical to Katz' cluster analysis on Latin American digitization,[44] with the only exception that the regression tree introduced an additional category with an average Internet usage of 86.68%. Based on his

---

[44]Katz, Raul et al. pp. 11.

cluster analysis, we could classify the first leaf node with an average share of Internet users of 13.53% as constrained countries, the second leaf node as emerging countries, the third one as transitional countries and the fourth and fifth leaf node as advanced countries.

6 Conclusion and Discussion

In addition to the analysis of general trends of Internet usage and the digital divide, the purpose of this study was to provide an understanding of the driving factors behind Internet usage at the macro level. For this, a variety of hypotheses were created and tested such as the impact of economic strength, free markets, access to electricity, urbanization, literacy, corruption and the level of democracy on Internet usage. In exploring these relationships, the empirical analysis of Internet usage consisted of three parts: the first part included an investigation of general time series trends and a ranking of countries according to their Internet usage. In the second part, scatterplot and correlation analysis were conducted and in the third part, a regression tree algorithm was applied, trying to predict Internet usage and to identify its explanatory variables. The empirical analysis provided evidence that the conventional view on the digital divide, in particular the importance of economic capital and material access, is the most important factor contributing to Internet usage. According to the regression tree, the second most important factor explaining Internet usage is the access to electricity, although this variable played a more important role at the global level than in Latin America.

Internet providing satellite systems have been once regarded as an alternative to other types of network access with the potential to close the digital divide. As this study showed, however, the market for Internet providing satellite systems is gradually disappearing with more and more developing and emerging countries catching up with more advanced societies in terms of Internet usage. In doing so, Latin American and Caribbean countries have followed a path similar to countries in the Asian-Pacific region and the Middle East. Moreover, countries in South Asia and Sub-Saharan Africa, too, slowly but gradually show an increasing share of Internet users despite massive population growth. As a result from these developments, satellite networks will probably remain a niche market for remote areas.

Christopher Yoon MSc. works as a researcher and data scientist at the Institute for Higher Military Leadership at the National Defence Academy in Vienna, where his academic attention focuses on cyberspace, digitization, data science and artificial intelligence at the operational level of war. After he graduated in political science and social science, he switched to economics and completed his master's in international relations at the University of Edinburgh. Before that, Christopher gained professional experience as research assistant at public universities and private think tanks, as an intern at the Austrian embassy in New Delhi and as an entrepreneur. His business activities focus on data analytics and the development of machine learning algorithms and applications for different sectors and industries.

Time to Change Your Education Programme—The Transformative Power of Digital Education

Alessia Durczok

Abstract

Changes in education are forcing us to adapt to new teaching methods and to discover new modes of studying. This development brings about positives sides to it and can be used to an advantage, for example, in the case of Latin America. The topics discussed in this chapter evolve around e-learning. It comes with advantages for some but may not always be the right mode of teaching, respectively learning. This article gives a rough guide to the planning of successful e-learning modules and a brief overview of what e-learning may look like. Furthermore, it discusses the example of nuclear security education and how e-learning changed the method of teaching. Latin America is dealing with challenges in qualitative tertiary education which e-learning can provide. Especially in the case of space applications, e-learning may serve as a potential way forward to spark interest in the topic or give space for 'learning by doing' in simulations.

1 Introduction

Flexibility is a buzz word in the world of today, along with professionalism, life-long learning, and innovation. We like to learn in a social context, when we watch others and copy them. An academic degree is supposed to get you a job and once you have that, you continue learning until you retire. What if you do not have access to education to get that job or the next promotion? Roughly 8% of the world's population lives in Latin America. Education there, has steadily been improving but it still characterized by the traditional classroom style, one to one

A. Durczok (✉)
Vienna, Austria

© Springer Nature Switzerland AG 2020
A. Froehlich (ed.), *Space Fostering Latin American Societies*, Southern Space Studies, https://doi.org/10.1007/978-3-030-38912-3_8

sessions and outdated teaching methods. Education is low in quality and dropout rates after secondary school are high, especially among women and minorities.[1]

Though we do not know much of what exactly is happening in the brain when we learn, we do know that learning is creating new pathways in the brain. Education is essentially a tool for change. It also has the power to change societies and speed up development. Considering Latin America's limited opportunities for tertiary education, this article will argue for the case of e-learning. E-learning as an emerging method of learning, gives flexibility, opportunity for micro learning and innovation. It may enhance your professional career or simply give you knowledge about of topic of your interest.

There is no recipe for the creation of the perfect e-learning module and it should be given careful consideration in the planning and development. The first part of this chapter will explain what e-learning means and a strategy on how to develop an e-learning module in six phases, according to the Certified European E-learning Manager (CELM) training, sponsored by the Lifelong Learning programme of the European Commission. These steps will support the successful development of new e-learning modules.

The second part of the chapter will shed light on the current situation in Latin America concerning education. Facts and figures underline the opportunity for the successfully implementation of e-learning in the region. Lastly, the case of nuclear security education illustrates the additional value e-learning may add for blended learning programmes. It may set an example for similar disciplines that are coined by similar characteristics, opportunities or challenges.

2   Transforming Education: What Is E-Learning?

In this article, the term e-learning is understood as various forms of digital media that support learning processes, where learning content is presented digitally. Examples include Web Based Training (WBT), Computer Based Training (CBT), E-lectures, explanatory Videos, mobile learning via smart phones or tablets, social learning, serious gaming or game based learning, virtual classroom with online lectures. Please note that Massive Open Online Courses, the so called MOOCS, are not the only form of e-learning.

Producing a successful e-learning programme is fairly straight forward if you have an e-learning manager, who is familiar with the complexity of organizing an e-learning module, and a clear picture of the content that should be conveyed. Though this may seem very easy and obvious to some readers, it is the fact that can make a project succeed or fail. E-learning currently has a very popular image and promises any organization or company to look modern. This article argues the case for e-learning but the reader should be aware of what that entails. E-learning is not a

---

[1]OECD (2015), "Executive Summary" in E-learning in Higher Education in Latin America, OECD Publishing, Paris.

set of slides which is posted online. E-learning has the power to change education. Though, it wants to be planned carefully.

2.1 A Rough Guide to Developing E-Learning Modules

The management of e-learning is a relatively new discipline in the field of education and training or human resource development. Because there has been a trend to develop modules 'as you go', the European Commission supported the project "Certified European E-Learning Manager"[2] (CELM) in an attempt to standardize the production of e-learning modules and thus setting a standard. The programme calls for the following six steps,[3] which shall be a guide to the planning of successful e-learning modules:

(a) a needs analysis,
(b) a concept design,
(c) a project plan,
(d) implementation,
(e) execution,
(f) an evaluation.

A needs analysis is the start of any education or training course. Understanding the needs of an organization, for example, will determine the type of course offered. Who needs to be educated? Who needs to be trained? Which subjects need to be taught? Which kind of qualification is needed for a successful outcome? Answering these questions will give the e-learning manager a better picture of which content the module shall entail and which instruction method is best. One should be aware, that this needs analysis may also show that e-learning may not be the best mode of study! The underlying idea remains that the module will create a link between knowledge that is present today and knowledge that should be present in the near future. If your goal is to increase the number of engineers working on space applications, the module will have to teach them engineering requirements and technical details of tools and machines needed for the production of equipment used for space applications. Consequently, if your goal is to spark interest for space applications, the module will focus on the use of space applications, for instance, in earth observation and the risk reduction of floods, tsunamis, fires, volcanoes and other natural disasters. Clear indicators that speak for e-learning, include that training material will be needed often, there are many learners that need to be trained over a long period of time, training material should be updated frequently, learners need to be able to study when and where they want, learning results need to be tested and show scores immediately. Should your analysis produce these indicators, you have good indicators that speak for e-learning.

---

[2]CELM (2019) Über CELM. http://www.c-el-m.de/v2/ueber-celm/ Accessed 13 Jul 2019.
[3]CELM (2019) Was. http://www.c-el-m.de/v2/was/ Accessed 13 Jul 2019.

During the conceptual design of the course, there are three main points, which need to be considered in the following order. Firstly, what shall be the didactical concept of the course? Should the learner study on their own or with a tutor? Will there be virtual lectures? Practical exercises are indispensable for quality courses. How will these be implemented? Secondly, what are the organizational limitations or requirements. What is the local learning culture? Who are the different individuals in the development of the course and what are their tasks? Which kind of training facilities are available? Thirdly, which kind of technology is feasible? Which technology is already available? Experience shows that very keen managers firstly acquire the technology and only later find out that it does not match their didactical concept. Hence, the indication to proceed in these steps. Answering all these questions above will also ensure that your resources are applied wisely.

Having completed the previous steps, will provide sufficient information to decide on a plan (which can be compared to project management plan). At this point, the e-learning project requires a detailed plan on the content of the course, the technology to be used, an action plan, a timeline, a list of resources, and a budget. The more detailed your planning is, the less difficulties you may encounter at a later stage. An often forgotten point at this stage, is the marketing of the module. No matter if you are planning e-learning for employees of an organization or a wider audience, marketing is an essential feature for success. Potential learners need to know that this e-learning module is being made available and which benefits it provides to them.

With all plans in hand, one can move up to the next stage which is the actual implementation of the module. This undertaking is of interdisciplinary nature, which means that stakeholder engagement is essential. Qualified trainers need to be involved, learners need to be informed, and technology should be tested. Experience of implementing previous e-learning modules showed that a test run of the learning management system is essential. Have user accounts been connected to the correct user role? If learners encounter technical difficulties, especially in the beginning, it will have an impact on their motivation and they may drop out of the course prior to completing it.

Theory is being translated into practice during the execution phase. A successful e-learning manager will keep in close contact with the learners to keep them motivated but also to manage change, as education creates change. This may mean that your enrollment capacity is limited, if there is only one tutor ho needs to be in touch with all students. Marketing continues during this phase. Learners need to know, how many hours a course will take, what the learning objectives are and what they get at the end of it. Learners also need to be made aware of which technical devices they are able to use.

Finally, the sixth phase is the evaluation of the module. It is important to find out what could have gone better. There is always room for improvement and there are many different tools which can be used for an evaluation. Learners should be asked for their evaluation but also the trainers. What do all the different stakeholders say? Has the time plan and budget been kept according to plan?

2.2 Benefits and Challenges

E-learning is a fairly new discipline since Elliot Masie coined the term in 1998. It has grown ever since and continues to refine its potential. A clear difference to the traditional classroom style method is that the focus switches to the learner. It is called a 'student-centric' view, that recognizes that not all students learn the same way. The flexibility of being able to use different digital media in one course and thus attempting to reach different learner styles is a distinct advantage of e-learning. Technology enabled learning can be of disadvantage when the appropriate technology is not available, it can however also be of benefit when the technology is available and long distances do not allow for a class with a teacher in front of the classroom. Technology and the internet are spreading all over the world. Latin America, as also other parts of the world, witnesses more new mobile phone connections than landlines being registered. Most mobile phone users are now working with smart phones. In the case of learning, it enables the learner to learn anytime and anywhere. It caters for different styles of learning. Formal or informal learning meets social learning with others, or mobile learning when you are on the go and most interestingly blended learning, which is a combination of various styles.

It catches the attention of the learner, because e-learning can make use of various media. Videos and images can supplement text. Stories, real case scenarios, case studies or even games make theory visible which is extremely helpful for visual learners. Audio files cater for aural learners, who respond particularly well to discussions or generally the spoken word. E-learning can even incorporate social media, which is ever present in our world, making use of the connections we already have. Simulations provide an inexpensive way to acquire information or an experience faster. They also offer an opportunity for sensitivity training. In the case of space studies, it could actually simulate a journey through space or a landing on the moon. These experiences can be simulated in a safe environment and allow for mistakes, which is a great opportunity to learn. Thus, e-learning transforms the way teaching and learning is taking place and also improves teaching and learning practices.

This rough guide for the production of an e-learning model is based on the European approach. Nevertheless, this can also be applied for the case of Latin America. The mere fact, that the internet does not know our State borders and easily overcomes long distances, means that almost anyone can take any e-learning anywhere they can access the internet. According to Hallberg, the United States, India, China and South Korea[4] are the top four countries in the world that have developed e-learning. Learners in Latin America can take these courses, apply their new knowledge locally to support the development of their continent. Of course, this argument is based on the assumptions, that the local government allows the website in question, that the broadband speed is strong enough to allow the student

---

[4]Adrian Hallberg (2017) The Top 4 Countries That Have Developed eLearning. https://elearningindustry.com/countries-that-have-developed-elearning-top-4 Accessed 13 Jul 2019.

to access the pages, and that the student is familiar with the language of instruction. E-learning may not be the perfect solution for all instances, though it offers many new opportunities that were not available in the past. Critics say that technology is limiting and not everyone has the opportunity to use a computer or have speedy internet. This criticism is correct. Nevertheless, e-learning is still able to reach a critical mass to make a difference. It is fair to use the opportunity if it makes a difference to some.

3 Tertiary Education in Latin America

In the beginning of 2018, Foreign Affairs complained about the higher education situation in Latin America and the Caribbean: "a deficit of graduates in science and engineering contributes to the region's low rate of innovation".[5] And that is exactly the point where educators and policies ought to start. Using this challenge as an opportunity in sparking interest in Science, Technology, Engineering and Mathematics, in short STEM, can build a stronger body of individuals who are interested in continuing education in more specific fields, such as space studies or related subjects. Paired with the circumstance that more and more space applications are increasingly being used in Latin America to support the continent's development, it is an ideal moment for students to enter the field with a positive outlook on actually finding employment after having completed their studies. As UN-SPIDER points out, space applications of the coming years will be a significant component of efforts in disaster risk reduction and the associated response and recovery mechanisms.[6] We may see an ever growing need on applications that help us solve the planet's problems of the future if we think about, for example, climate change, natural disasters, or pollution.

Higher education is a key component for development and thus a significant factor for Latin America. According to UNESCO, tertiary education has expanded rapidly in the region since the early 2000s[7] and the OECD admits that the rise of e-learning can have a positive impact,[8] though inequality persists. The interesting fact is that, though many may not have access to university education, they seem to have access to the internet. The World bank suggests that 98% of people in Latin America have access to a mobile phone,[9] which offers a wide range of new

---

[5]Foreign Affairs (2018) At a Crossroads: Higher Education in Latin America and the Caribbean https://www.foreignaffairs.com/reviews/capsule-review/2017-12-12/crossroads-higher-education-latin-america-and-caribbean Accessed 13 Jul 2019.

[6]UNOOSA (2019) UN SPIDER Knowledge Portal: Space Application http://www.un-spider.org/space-application Accessed 13 Jul 2019.

[7]UNESCO, EFA Global Monitoring Report, https://en.unesco.org/gem-report/sites/gem-report/files/178428e.pdf Accessed 13 Jul 2019.

[8]OECD (2015), "Executive Summary" in E-learning in Higher Education in Latin America, OECD Publishing, Paris.

[9]Latin America leads Global Mobile Growth http://www.worldbank.org/en/news/feature/2012/07/18/america-latina-telefonos-celulares Accessed 13 Jul 2019.

opportunities to reach out to students but also for students to reach educational programmes that may be out of their reach. The OECD report further laments that most e-learning courses in the region are not accredited[10] and thus lack a quality assurance mechanism. Furthermore, there is a gap between the skills that are being taught versus the skills in demand by the industry.[11] Though, this being an understandable problem, it can easily be rectified by a thorough gap analysis during the needs analysis phase as described above. Lastly, modern learning and teaching methods have hardly been introduced into the region[12] and there seems to be a reluctance to adopt these. Again, this is an issue which can easily be changed with e-learning and deserves thorough discussion during the conceptual design phase as explained previously. Nonetheless, the OECD report concludes that "it is relevant to increase the availability"[13] of e-learning in the region and it seems to be a real opportunity despite of the existing problems.

4 Case Study: Why the IAEA Nuclear Security Education Programme Introduced E-Learning

Only several years ago, the topic of nuclear security education underwent a transformative change with the development of e-learning courses. The following case presents an example where e-learning has been used in several modes to either raise awareness, reach out to learners who could not attend training courses, provide a reference for technical equipment anytime and anywhere, spread basic knowledge free to anyone with an interest in the topics. The International Atomic Energy Agency (IAEA), has made all their e-learning modules available online and accessible to any user.[14] At the time of writing the article, 16 different modules in four nuclear security disciplines appeared on the webpage.

The increasing demand of nuclear energy and nuclear applications, for instance, as cancer therapy in medicine, in agriculture and other industrial uses, have also augmented the need for qualified nuclear security professionals. This includes policy makers, lawyers, scientists, border control but also technical experts, security guards and many more. As it is the State's responsibility to ensure security of nuclear or radioactive material, it was also the Member States of the International Atomic Energy Agency (IAEA), who decided for an extensive nuclear security education and training plan in 2005[15] at the Board of Governors meeting in Vienna.

---

[10]OECD (2015), "Executive Summary" in E-learning in Higher Education in Latin America, OECD Publishing, Paris.
[11]Ibid.
[12]Ibid.
[13]Ibid.
[14]International Atomic Energy Agency (2019) Nuclear Security E-Learning https://www.iaea.org/topics/security-of-nuclear-and-other-radioactive-material/nuclear-security-e-learning Accessed 13 Jul 2019.
[15]International Atomic Energy Agency Educational Programme in Nuclear Security, IAEA Nuclear Security Series No. 12, IAEA, Vienna (2010).

The need for human resource development programmes, sparked the development of a curriculum for a certificate or master programme, illustrated in the *IAEA Nuclear Security Series No. 12, Educational Programme in Nuclear Security*[16] *(NSS12)*. This programme was available and free to use for any Member State to support appropriate education and training mechanisms locally for any organization that has to do with nuclear security.

In 2005, there were only a few universities worldwide which taught nuclear security topics as part of a wider programme for safety or safeguards studies. The new publication NSS 12, enabled academic institutions worldwide to quickly add a new curriculum in nuclear security to their programme, with an international focus which had been written by international experts and was approved by the international community. Additionally, the IAEA facilitated the International Nuclear Security Education Network (INSEN), which gave universities with an interest in nuclear security, a forum to exchange ideas, discuss international differences in education and potentially even exchange expertise.

INSEN was born in 2010 and 9 years later, it had more than 170 academic members, who collaborated on different projects. Most significantly, they came together to develop teaching material, textbooks, lecture notes, and composed a book of real case studies, that can be used for teaching. The curriculum of NSS 12 explained 22 subtopics of nuclear security in detail. INSEN provided material for each topic, which was written and approved by international experts.

Universities had to adjust this programme to their local needs. A Masters or a certificate degree, however, from a student's perspective is a timely and also financial investment. Only a few students were keen to onboard these programmes and surveys would show that they either did not know enough about nuclear security to spark their interest, or they did not know about the availability of these programmes.

In order to spark interest for deeper studies and the desire for a programme that offered a quick start and a broad overview to nuclear security, the IAEA initiated the International Nuclear Security School, which annually takes place at the International Centre for Theoretical Physics (ICTP) in Italy. This two week intensive programme, allowed students with various backgrounds in a discipline of relevance to nuclear security,[17] to apply for the course. Each theoretical concept taught in the class was accompanied by practical exercises and complemented by a technical visit to a nearby port, which illustrated the newly acquired knowledge in practice.

Additionally, the above mentioned 16 e-learning modules in the various sub-categories of nuclear security supported the spread of basic nuclear security knowledge by guiding the student through the topic and testing knowledge via a quiz. Applicants to the nuclear security school were only admitted to the course, if they had completed all available online modules. This had two advantages. Firstly,

---

[16]Ibid.

[17]ICTP (2019) Joint ICTP-IAEA 2019 International School on Nuclear Security http://indico.ictp.it/event/8650/ Accessed 13 Jul 2019.

in some cases it was the first time that the student became familiar with the subject matter and could do so it their own pace. Important terminology was projected on the screen, as visual support, and useful to English-second-language speakers. Secondly, the e-learning module prior to the start of the class, ensured that all students started with the same basic knowledge. This would help the teacher to plan the class better to achieve the learning objectives of the class. Specifically, for training (rather than education) some e-learning modules became part of the course as blended learning. For the overall education programme, the introduction of e-learning modules provided a necessary bridge between the education and training programme and offered a quick solution to interested learners at their own pace.

4.1 Did It Work?

Students who attended the school were able to return to their home country and start working on a nuclear security infrastructure. Of course, deeper studies or expert knowledge in specific subjects are needed to successfully implement nuclear security mechanisms but it gave a quick start to, for instance, policy makers, who could then gather a team of experts and make educated decisions on the topic. Some candidates who attended the school, then also continued their education by enrolling in one of the Masters programmes in nuclear security. Different options of full time studying or long distance learning, completed the wide ranging offer in educational programmes available to the international community.

Space studies are of course focusing on different topics than nuclear security. Both subjects, however, conform in their necessity for a technical understanding and a great amount on tacit knowledge. Both subjects are neither just science or technical. Their interdisciplinary nature is what makes it complex and since neither belong to the traditional subjects a university offers; they are also rare to find. This of course hinders the appropriate development of human resources and will undermine the long term success and the sustainability of any space programme.

5 Conclusion

E-learning in general is a young discipline, which keeps on developing, and can present itself in various different formats. This article touches upon the surface of what e-learning is, what it can be and how it transforms education in order to support development in Latin America. The general trend in education is moving away from the traditional classroom setting, to offer new didactical methods and hopefully better learning environments for the learner. A well planned and thought through e-learning module can make a difference to higher education. Various different digital media provide a motivating environment to the learner accompanied by different modes of studying, as different learners come with different needs and or interests.

For the development of a qualitative module, one should consider the six basic steps, which CELM suggest, namely a needs analysis, concept design, project plan, implementation, execution, and evaluation. E-learning offers various new opportunities which did not exist in the past but there may be cases where e-learning is not appropriate. It is significant to note that the focus is moving away from the classroom and towards the learner. One size fits all, is an outdated concept in education and e-learning is able to recognize such.

Though education in Latin America is steadily improving, there is room and an excellent opportunity for e-learning. The case of nuclear security has shown that an international programme must give flexibility to the learner and that e-learning is not necessarily a standalone course. Blended learning is a great alternative and closed a gap between education and training. This strategy can of course also be applied to other disciplines and education programmes. A thorough needs analysis will show the value of e-learning to any other programme.

Alessia Durczok holds a Master in Advanced International Studies from the Diplomatic Academy of Vienna, Austria. She used to manage the Nuclear Security Education programme at the Division of Nuclear Security at the International Atomic Energy Agency, where she was part of the team that introduced e-learning to the programme. She received the CELM certification while researching education and e-learning methods and is eager to find out more about the changes of education in the brain.

Changing Objectives in Achieving Space Sovereignty in Latin America

Mehak Sarang

Abstract

The history of space activities in Latin America is defined by a quest for sovereignty beginning in the early 80s with the desire to develop a regional satellite as a challenge to the Western foothold in the telecommunications sector. Ultimately, the region failed to bring this dream to fruition; instead, independent countries found more success in developing their own satellites. More recently, emerging nations, such as Peru, have found similar success, largely due to recent technological developments in the sector and increased reliance on international partners. What led to success in these efforts, as opposed to failure in earlier ones? One factor may be the original construction of the satellite as an anti-hegemonic symbol, which led to the inability of the region to successfully transform political will into space assets. This article examines the shift away from the definition of "sovereignty" in space as solely ownership of space assets, arguing that this has played a role in the success of nations in the region in pursuing effective space policy that emphasizes the benefits of space applications in improving the lives of everyday citizens.

1   Introduction

The list of Latin American countries with satellites is constantly growing. Countries such as Venezuela, Argentina, and Brazil led the way in the early 80s by acquiring and developing satellite technology. Other countries in the region, although historically barred by cost, long expressed their desire to establish satellite sovereignty.

M. Sarang (✉)
Wellesley College, Wellesley, MA, USA

© Springer Nature Switzerland AG 2020
A. Froehlich (ed.), *Space Fostering Latin American Societies*, Southern Space
Studies, https://doi.org/10.1007/978-3-030-38912-3_9

Currently, whether via ownership, operation, or in-house development, satellites still provide a clear first step for emerging nations towards becoming space-faring.[1]

Emerging nations seeking a path towards establishing a presence in orbit may have a variety of motivations, but the main focus of policymakers in Latin America tends to be to advocate for space as a tool for development and an area of international cooperation. As demonstrated by more established space nations in the region and elsewhere, significant socioeconomic benefits derived from space are attractive to emerging countries. Beyond providing a whole host of downstream applications, the ability to construct, own, or operate space technology is indicative of the level of economic development and the health of the innovation system of a country. Furthermore, seeking international partnerships allows countries, via technology transfer and capacity building programs, to capitalize on more advanced nations' expertise to accelerate development.

However, another strong political motivation in the region is the desire for Latin American countries to exert their independence via ownership, operation, or development of proprietary space assets. As discussed by Lisa Park, "satellite footprints enable satellite operators, companies, and nation-states to forge new political alliances and trade relations, establish new forms of capital and exchange, and develop new forms of social control…[and] function as counter-hegemonic projects."[2] Satellites are used by telecommunications companies, multinational corporations, the global media, and governments for data transmission, telephone traffic, television and radio broadcasts, as well as military purposes. However, for years, satellite production and use was concentrated in the Global West, forcing other countries to rely on information streams filtered through Western (mostly U. S.) satellites. In Latin America, establishing independence in this realm was of utmost importance, as most countries were emerging from years of U.S. influence in their socio-political spheres.

On the whole, these motivations have remained relatively stable when one looks at the history of the development of satellites throughout the region, but the space industry of 2019 is vastly different to that of 1980. The cost of developing satellites has been reduced significantly with advances in CubeSat and launch technologies. Concurrently, with the establishment of more space-faring nations in non-Western countries, emerging players have access to new, non-traditional international partners. These new factors have influenced the modes of collaboration and motivations regarding satellite acquisition in the region, especially as the number of Latin American countries with space-faring capabilities continues to grow. Most notably, what constitutes "satellite sovereignty" has fundamentally changed since the early 80s, when development of a regional satellite in Geostationary Orbit (GEO) was the main focus on the continent.

---

[1]Wood, Danielle. (2012). "Charting the evolution of satellite programs in developing countries— The Space Technology Ladder." Space Policy. 28:1, 15–24, DOI: https://doi.org/10.1016/j. spacepol.2011.11.001.
[2]Parks, Lisa. (2012). "Footprints of the Global South: RascomQAF/1R and Venesat-1 as Counter-hegemonic Satellites," in Handbook of Global Media Research. Ingrid Volkmer, ed. Wiley Blackwell, 2012, 123–142.

Although the desire for a regional Latin American satellite, and in turn, sovereignty, was quite strong, due to a series of missteps, political will could not be transformed into achieving this objective. By tracing new developments in the current industry and new modes of satellite acquisition, it will be clear that the success of emergent countries in building and operating proprietary satellites is due to a shift in the idea of sovereignty in space.

2 Seeking a Regional Satellite: Historical Perspectives

Still reeling from the effects of years under U.S. hegemony, Latin American countries were determined to strengthen regional ties in the period following the Cold War. These intentions were reflected in the landmark Cartagena Agreement of 1969. As stated in the agreement, the Andean governments of Bolivia, Colombia, Ecuador, Peru, and Venezuela resolved to

> strengthen the union of their peoples and to lay the foundations for advancing toward the formation of an Andean subregional community; aware that integration constitutes a historical, political, economic, social, and cultural mandate for their countries, in order to preserve their sovereignty and independence.

Issues of regional sovereignty and independence were firmly in the background as Andean countries set out to establish a joint space policy and strategy. Eager to avoid the development of yet another Western hegemonic world order in space, the 1976 Bogota Declaration was an early attempt by a group of equatorial nation-states (Colombia, Congo, Ecuador, Indonesia, Kenya, Uganda, Zaire, and Brazil) to claim sovereignty over the geostationary orbital slots above their national territories. They argued that the "first-come, first-serve" nature of orbital allocation by the International Telecommunications Union (ITU) was unfair to developing nations that were unable to launch assets to occupy the valuable slots superjacent to their territory.[3]

Although ultimately not legally authoritative, the declaration was a demonstration of the anti-imperialist sentiments of countries—they would not watch from the sidelines while the developed world sought to dominate outer space. Intelsat's clear monopoly in the global telecommunications sector was just one demonstration of the growing "footprint" of Western satellites. At the International Telecommunication Convention, contracting countries declared that they "...fully recogniz[e] the sovereigns right of each country to regulate its telecommunication for the preservation of peace and the social and economic development of all countries..." Providing domestic and regional programming separate from that which was provided by Intelsat, became a priority. Initially, challenging this monopoly seemed like an insurmountable task. However, when Intelsat ultimately began to cede international control as private carriers and fibre optics gained traction, other

---

[3]Durrani, Haris. "The Bogotá Declaration: A Case Study on Sovereignty, Empire, and the Commons in Outer Space." Columbia Journal of Transnational Law, http://jtl.columbia.edu/the-bogota-declaration-a-case-study-on-sovereignty-empire-and-the-commons-in-outer-space/.

competitors saw an opportunity.[4] As described by Pía Riggirozzi in *The Rise of Post-Hegemonic Regionalism: The Case of Latin America,* it would be because of the "collapse of U.S.-led hemispheric leadership" that Latin America would be compelled (or required) to develop "alternative institutional structures and cooperation projects."[5] Whether it was due to the decline of Intelsat, the desire to "bust open" the Intelsat monopoly (as was the aim of Pan-American Satellite Corporation founder Reynold Anselmo), or the rush to place a satellite in orbit to reassert the failed Bogota Declaration, it soon became clear that the next natural step towards asserting autonomy would be through developing a regional satellite.[6]

In hopes of developing a regional satellite, members of the Andean Telecommunications Enterprise Association (ASETA) contracted Canada Astronautics Limited/satTel Consultants to carry out the first feasibility study for the development of three 12-transponder satellites to be launched by 1982 or 1983. The study was deemed "Project Condor," but ultimately failed to gain traction among the member countries due to financial limitations. The feasibility studies failed to address the funding of the project, as well as how ASETA members would manage operation of the regional satellite or its ultimate purpose.[7] Even after the creation of a separate committee imbued with the sole purpose of carrying out Project Condor, the project never came to fruition. Alternatives were considered, but when Intelsat offered a shared lease agreement in 1984, ASETA members refused the offer, reasserting the importance of political independence from external partners.[8]

Driven by a top-down mentality and relying on foreign experts with little understanding of the needs of the region proved to be fruitless. Yet, this pattern continued—as demonstrated by the next attempt at acquiring a regional satellite by the Andean Community (CAN). The CAN was established in the 70s by member states Bolivia, Colombia, Ecuador, and Peru in order to strengthen political, social, economic, and cultural integration. With the creation of the Andean Committee of Telecommunications Authorities (CAATEL) in 1991, CAN aimed to encourage investment in telecommunications services by eliminating trade barriers and adopting other measures, ultimately hoping to define a common space policy. Concurrently, the ITU awarded CAN a GEO orbital slot in orbital position 67° West, providing the community with the opportunity to launch a sovereign satellite to achieve some of these goals. After being awarded this slot, CAN adopted a resolution that restricted use of the slot to a company from the region. However, according to ITU regulations, if the Andean Community did not provide

---

[4]Payne, S. (1988). "Earth to Intelsat: The Party's Over." Business Week 5 September, p. 61.

[5]Riggirozzi, Pia & Tussie, Diana. (2012). The Rise of Post-Hegemonic Regionalism. https://doi.org/10.1007/978-94-007-2694-9.

[6]Shearmu, Richard, Carrincazeaux, Christophe and Doloreux, David, (2016), Handbook on the Geographies of Innovation, Edward Elgar Publishing.

[7]Ospina, S. (1988). "Project Condor: An Analysis of the Feasibility of a Regional Satellite System for the *Andean Pact Countries*." Unpublished Dissertation. Montreal: McGill University.

[8]Jakhu, Ram & Pelton, Joseph. (2017). "Global Space Governance: An International Study." Space and Society. Springer, 2017.

documentation with the intention to place a satellite in the orbit, the ITU would free the position for use by other countries.[9]

One of the earliest attempts to fill the slot was to authorize the development of the Andean Satellite System Simon Bolivar (ANDESAT) in 1998. Ultimately, this partnership failed, and the authorization was withdrawn in 2005. Following this failure, CAN authorized the Bolivarian Republic of Venezuela to place a temporary satellite in the orbit. Unfortunately, this was yet another partnership that never came to fruition due to policy differences and the ultimate decision of Venezuela to exit the community. In November 2006, facing a looming deadline and the possibility of losing the slot, CAN adopted a resolution allowing international commercial use of the spectrum allowing the community to contract companies from anywhere in the world. In 2014, CAN signed a contract with SES NEW SKIES B.V., a telecommunications and satellite company from the Netherlands.[10] In return for filling the slot, the company was granted access to the satellite as well, meaning CAN would not be able to fully exploit the mission objectives as they would not have full ownership over the orbital position.

In the end, the early regional effort to develop a regional satellite was never fully realized. Driven by a top-down approach, the motivations expressed by member states were never successfully translated to action. Bogged down by financing issues and policy coordination, regional cooperation proved to be difficult. Furthermore, when international partners were consulted, the results were not collaborative in the ways that the countries had hoped. In the push for satellite sovereignty as a challenge to Western hegemony, the function of the satellite itself, policy coordination among regional members, and financing were left as afterthoughts.

3 Learning from the Past: New Modes of Satellite Acquisition

After the failure of the dream of a regional satellite, independent nations began to develop or obtain their own satellites, reflecting new modes of satellite acquisition, motivations, and modes of collaboration. Recently, countries in the region without a history of space activities have also expressed interest in developing or purchasing satellites. Modern approaches to satellite acquisition stand in direct contrast to the series of missteps encountered by the Andean Community in their quest for a regional satellite. New trends in the region are exemplified by the development of PeruSat-1, the Peruvian satellite launched in 2016, especially when contrasted with the historical effort.

---

[9]Guzmán Gómez, Camilo & Camila Iannini, Maria. (2013). "The Andean Community Failure to Create a Common Space Policy." Paper presented at the International Astronomical Congress in Beijing, China.
[10]SatelliteToday.com (2015) SES Training Andean Community on Satellite Communications, accessed 5/30/2019, https://www.satellitetoday.com/telecom/2015/09/22/ses-training-andean-community-on-satellite-communications/.

PeruSat-1 is Peru's first Earth observation optical satellite system. Constructed by Airbus Defence and Space, with the support of the French government, the satellite was launched on September 15, 2016. Peru had long expressed interest in obtaining a proprietary satellite, and was a crucial advocate for satellite sovereignty in the history described above. However, the process of acquiring PeruSat-1, mission objectives, and the selection of Airbus as the sole prime contractor are very telling of the ways in which satellite development and acquisition have evolved since the 80s. Namely, two of the main goals of the mission were to promote knowledge-sharing and capacity building and the downstream applications that would support the development of Peru: with this shift in focus, sovereignty in space took on an entirely new meaning.

3.1 Capacity Development and Knowledge-Transfer

When choosing a contractor for the mission, Airbus' proposal was, controversially, not the most economical. However, two aspects put the proposal over the edge: the establishment of a government-to-government partnership between Peru and France, and Airbus' commitment to training 40–50 Peruvian engineers during the course of satellite development.[11] In Toulouse, France, Peruvian engineers were trained in elements of satellite construction and image analysis. In Lima, Airbus developed an image analysis and processing center, where engineers were trained using Airbus' current Earth observation satellites. Airbus CEO Eric Béranger said of the training program:

> Of course, the nations want their own satellite systems...but also a way to develop a sovereign space industry and competencies through technology transfers. In the case of PeruSat-1, this includes a significant knowledge and know-how transfer, but it also includes a satellite simulator and specific application fields based on space optical imagery.

As shown through the historical analysis, the focus of early sovereignty in space was proving the ability to occupy orbital space. Alternatively, in the case of PeruSat-1, the development of the satellite itself served as a means to an end. One of the main objectives of the mission was to increase space industry competencies to ensure sustainable sovereignty. True independence would be realized when reliance on outside partners was no longer necessary. Thus, contracting was not a preferred method of international cooperation because the long-term government-to-government partnership provided more opportunities for capacity development and technology transfer.

The rise of CubeSats on the continent reflect this trend. A growing number of universities in the region are launching their own programs to construct CubeSats in-house, relying on international agreements (largely with countries such as China, India, and Japan) to launch or operate the satellites. In Costa Rica, the Central

---

[11]Space News (2015) Vega to Launch Peruvian Imaging Satellite Along with Skybox Craft, accessed 5/28/2019, https://spacenews.com/vega-to-launch-peruvian-imaging-satellite-along-with-skybox-craft/.

American Association for Aeronautics and Space (ACAE), which was founded in 2010, announced the development of the first domestically-built satellite. The 1-unit CubeSat was deemed Project Irazú, and launched in February 2018 to measure atmospheric carbon dioxide levels. SpaceX launched the satellite and, via an agreement with the Kyushu Institute of Technology in Japan, the satellite itself was deployed via JAXA's Kibo module. In a brilliant display of the "democratization" of the space sector, $75,000 of the total cost of the satellite ($500,000) was raised via a crowd-sourcing campaign. According to the crowd-funding website, the emphasis of the mission was on capacity building—more than anything, the satellite demonstrated that Costa Rica could build indigenous technologies.[12] Although the project personnel and various advisors were international, the core team of engineers was comprised of Costa Rican citizens. This was crucial to the mission because, as described on the website, "one of the most efficient ways to promote space technology development in Costa Rica is to prove its capacity to develop a full-scale space engineering project involving a scientific mission related to one of the country's needs." The Irazú mission, much like PeruSat-1 and numerous other CubeSat missions in countries such as Ecuador, Uruguay, and Guatemala, represented the ability of Latin American countries to independently meet development goals—thus redefining the notion of sovereignty.

3.2 Space Data as a Means for Development

Beyond independent development of technology, sovereignty also came to be defined as the ability to acquire and use data to meet development objectives. After becoming fully operational in December 2016, PeruSat-1 has acquired more than 102,200 images used by more than 70 public entities, as reported in March 2018. The satellite also allows Peru to share data; and the country recently signed an agreement with South Korea, with deals proceeding in Azerbaijan, Kazakhstan, Argentina, and Brazil. According to Gustavo Henriquez, head of the National Commission for Aerospace Research and Development in Peru (CONIDA), "You can use images for agriculture with crop monitoring, for the supervision and safety of our cultural heritage and for the prevention and monitoring of illegal activities such as illegal mining, deforestation and detection of clandestine airstrips." After launch, the emphasis has been on highlighting data production, and promoting the utility of that data.

Colombia's struggle to develop a national space policy is a tale that spans decades but ultimately demonstrates the shift to prioritizing data acquisition and production. By the end of the seventies, Colombia's president at the time, Julio Cesar Turbay, began the acquisition process of a telecommunications satellite that was halted by the following president, Belisario Betancourt. SATCOL was another attempt at satellite acquisition that was brought to an end due to concern that US

---

[12]Kickstarter (2016), Irazú Project: The First Satellite Made in Costa Rica, accessed 5/28/2019, https://www.kickstarter.com/projects/irazu/irazu-project-the-first-satellite-made-in-costa-ri.

corporations were exerting undue pressure throughout the project. More recently, the Colombia Presidential Program for Space Development was established in 2013 to encourage universities and companies to enroll in space development in the country and made strides towards acquiring a remote sensing satellite.[13] After one year, the program was cancelled. The decision was justified by the government by arguing that purchasing remote sensing data would be cheaper. Thus, the important factor in developing a space policy was not simply ownership of space assets but strategically considering the outcomes and applications of space data. Colombia's Spatial Data Program25 is a successful example of these shifting objectives. The program manages the use of information coming from space systems, mainly satellites. The data is unified and systematized so that governmental institutions can avoid redundancy in data acquisition and be more efficient in data use.[14] Therefore, in a way, this limited success can be lauded in that the government was able to recognize the importance of coordination and access to space data. Data can be acquired without the ownership or operation of physical assets, once again reflecting a shift in space policy objectives, especially in achieving "sovereignty."

4 Conclusion

For developing nations around the world, emerging from a colonial past can often lead to a desire to seek sovereignty in many aspects of the country's government and economy. In Latin America, the quest to obtain a regional satellite became a symbol of regionalism in defiance of Western hegemony. However, this motivation was ultimately not powerful enough to fully realize the objective. Unable to secure financing and properly manage regional politics, the Andean countries in particular failed to organize development of a fully independent "sovereign" regional satellite.

More recently, nations have found success in developing domestic satellite development capabilities, or found ways to benefit from the applications of space data. The quest for sovereignty still plays an important role in these programs; but sovereignty is defined and achieved through alternate means. Partnership with long-term actors and institutions is seen a boon, as it provides technology transfer and capacity-building opportunities. Applications of satellite technology in aiding development, rather than the satellite itself, have become the focus of missions.

More broadly, the change in the policy of the region can be characterized as shifting from a top-down, collective approach to embracing a bottom-up, independent approach. As stated by Luis Enrique Salaverría, the President of the El Salvador Aerospace Institute, at the 3rd International Space Forum of 2018 held in Buenos Aíres,

---

[13]Guzman Gomez, Camilo, The Creation of a Space Policy in Colombia, a Chaotic History? (October 1, 2012). Proceedings of the International Astronautical Congress (IAC-12), 2012 ISSN:0074-1795.
[14]Becerra, J. (2014). Colombia's space policy: An analysis of six years of progress and challenges,. Acta Astronautica, 100, 94–100. https://doi.org/10.1016/j.actaastro.2014.03.018.

From my experience in my country, I can tell you that things do not usually work from top to bottom. What I mean to say is that private citizens or enterprises are the first to become interested in exploring, understanding and using space technology and only after they are successful is that the government becomes interested and actually gets involved in creating policies to support and promote these activities. Therefore, a good idea is to identify and support those early actors, such that they can truly make a difference in their respective countries.

Originally, space was the theater where questions of global order were played out. Satellites, and their resulting "footprints," were an indication of power. Latin American countries disputed, and ultimately wished to occupy, orbital space as a counter to the growing Western presence. While these issues remain relevant, especially in issues of space as it relates to security, the growing emphasis has been on utilizing space for development. Now, Latin American government seek ways to support space applications. Thus, the notion of sovereignty, once symbolized by the ownership of assets and occupation of territorial space, has evolved. Instead, the ability of emerging countries to independently utilize space for their needs is the true measure of sovereignty and ultimately, has allowed even emerging space nations in the region to translate intention into action.

Mehak Sarang received her B.S. (2018) in Physics from Wellesley College in Wellesley, MA. She has been involved with research regarding the rise of NewSpace, and the role of the government in providing support to private industry. Currently on the Susan Rappaport Knafel '52 Traveling Fellowship, she is studying the space industry in seven countries, interested in the ways in which individuals will benefit from space exploration.

Long-Term Sustainability of Space Activities: Achievements and Prospects

Laura Jamschon Mac Garry

Abstract

This article aims to highlight the main areas of discussion regarding long-term sustainability of space activities within the United Nations and assess the most significant challenges COPUOS has towards the establishment of a new working group to make progress in the field. To this end, it will account for the evolution of the negotiations on sustainability, focusing particularly on the proposal made by the Group of Latin American States and the Caribbean (GRULAC). In addition, this article will briefly outline the European initiative on a code of conduct on space activities and will contrast the projected text with the guidelines on long-term sustainability, agreed upon at the Committee on the Peaceful Uses of Outer Space (COPUOS). Finally, it will argue that while certain aspects of space safety and security are addressed in other bodies, these aspects are a fundamental premise of the sustainability of activities in outer space.

1 Introduction

Dependence of our societies on space activities has become an indisputable fact. Today satellite technology provides hundreds of services that are essential for the normal functioning of a State, such as meteorological services, monitoring of areas affected by natural disasters, the provision of communication systems, military reconnaissance and surveillance, just to name a few.

This article is written in the author's personal capacity. The opinions expressed here are the author's own and do not necessarily reflect the position of the Argentine Republic.

L. Jamschon Mac Garry (✉)
Università La Sapienza, Rome, Italy
e-mail: laura.jamschonmacgarry@uniroma1.it

© Springer Nature Switzerland AG 2020
A. Froehlich (ed.), *Space Fostering Latin American Societies*, Southern Space
Studies, https://doi.org/10.1007/978-3-030-38912-3_10

The risk of such services being affected is not so remote. The causes may lie in both natural phenomena (such as space weather and the impact of asteroids) and human action (kinetic, electromagnetic or cyber attacks). Another critical threat is the increasing number of space debris, which is even more concerning in the current reality, where small satellites and large constellations proliferate.

Therefore, security, safety and long-term sustainability of outer space activities are vital issues that the international community should address through a comprehensive strategy. They touch on matters that somehow fall under the mandate of different multilateral bodies of the United Nations system, as COPUOS, the Conference on Disarmament and the First and the Fourth Committees of the General Assembly. Therefore, the international community faces the challenge of striking a balance between addressing the subject holistically, avoiding duplication of work in different fora and maintaining full consistency.

This article aims to make a brief review of the work on security, safety and sustainability of outer space. To this end, it will examine the discussions within COPUOS for negotiating a set of guidelines on the long-term sustainability of space activities. Second, it will outline the proposed Code of Conduct of the European Union, and other initiatives carried out within the framework of the United Nations also related to the safety, security and sustainability of outer space activities. With these elements, it will assess the progress made, the steps and challenges ahead within the framework of COPUOS, arguing that safety and security are a fundamental premise of the long-term sustainability of space activities.

2   Sustainability on the Agenda of COPUOS: A Decade of Work

At the outset, it is necessary to make some preliminary clarifications of the terminology to be used throughout this chapter. Sustainability of space activities is closely linked to space security and space safety. While the former refers to threats caused voluntarily and of an aggressive nature, the latter deals with non-voluntary threats.[1] It is for this reason that some authors use the abbreviation '3S' (safety, sustainability and security) to refer to all of these concepts in a unified and interrelated manner.[2]

In turn, space security tends to encompass three different meanings: the use of space objects for security and military objectives (outer space for security), security of space objects against risks and natural or man-made hazards (security in outer space) and safety of people and the environment on Earth against natural disasters

---

[1]See Jana Robinson, "The Status and Future Evolution of Transparency and Confidence-Building Measures", ESPI Report No 27, September 2010, 10; Sergio Marchisio, "Security in Space: Issues at Stake", *Space Policy* (2015), 1.

[2]See Peter Martinez, "Challenges for ensuring the security, safety and sustainability of outer space activities", *Journal of Space Safety Engineering* 6, no 2 (2019).

and risks from outer space (security from outer space).[3] When this article speaks about the sustainability of space activities, it refers to the second meaning.

In 2009, the French delegation proposed including an agenda item on the long-term sustainability at the Scientific and Technical Subcommittee (STSC) of COPUOS.[4] This new agenda item entitled 'Long-term Sustainability of Space Activities' was mainly focused on space debris and its impact on space traffic.[5] It is worth noting that the French delegate Gérard Brachet had already proposed the topic during his Chairmanship of COPUOS (2006–2007).[6] However, the issue at that time had not acquired the necessary maturity for treatment.

The original proponents of the topic had suggested a bottom-up approach. This method required that discussions be initiated at a technical level so that when the document was submitted to the decision-making level, it would be easier to achieve agreements, avoiding reopening the negotiated text.[7] A clear example of such an approach are the Space Debris Mitigation Guidelines of the Committee on the Peaceful Uses of Outer Space, one of the instruments of soft-law that made a significant contribution to security in outer space and sustainability of space activities.

From the beginning, the guiding idea of negotiating a set of guidelines on the long-term sustainability of space activities was not to substitute or amend the international legal framework of outer space but to complement it with a non-binding instrument that would take into account the new challenges in the space field. That endeavour would aim to promote monitoring, communication and international cooperation in order to avoid future collisions, interference and disruption of satellite information and safeguard the regular operation of space missions.

At that time, an informal group composed of States interested in the subject started to meet with the Office of Outer Space Affairs (OOSA) and the European Space Agency (ESA). The group focused on threats and natural causes of disturbances affecting space systems (space weather, solar eruptions, micrometeorites, for instance).[8] On those days, proponents were not in a position to include more current issues, such as 'jamming', 'spoofing' or cyber attacks. On the contrary, COPUOS had already been dealing with an agenda item on space debris since 1994,[9]

---

[3]Jean Francois Mayence, "Space Security: Transatlantic Approach to Space Governance", ESPI Report No 27, September 2010, 35. See also: Michael Sheehan, "Defining Space Security", in *Handbook of Space Security. Policies, Applications and Programs*, ed. Kai Uwe Schrogl, Peter L. Hays, Jana Robinson, Denis Moura, Christina Giannopapa, (New York: Springer, 2015), 8 and 10.
[4]Report of the 52nd Session of COPUOS (2009), UN Doc A/64/20, para. 161.
[5]Report of the 46th Session of the STSC (2009), UN Doc A/AC.105/933, para. 80.
[6]Future role and activities of COPUOS (submitted by the Chair), UN Doc A/AC.105/L.268, 10 May 2007 and UN Doc A/AC.105/L.268 Corr. 1, 1 June 2007.
[7]Gérard Brachet, "The Origins of the Long-Term Sustainability of Outer Space Activities", *Space Policy* 28, no 3 (2012): 162.
[8]Long-term sustainability of activities in outer space (submitted by France), UN Doc A/AC.105/C.1/L.303, 9 February 2010, para. 13.
[9]Report of the 31st Session of the STSC (1994), UN Doc A/AC.105/571, paras. 63–74.

an effort that reached its highest point with the adoption of the aforementioned Space Debris Mitigation Guidelines in 2007, fifty years after the launch of Sputnik 1.

This preliminary informal work tried to mature the issue for its inclusion on the agenda of COPUOS, which was delayed until 2010. It can be argued that the trigger that ended up precipitating its inclusion was the accidental collision of the satellite Iridium 33 and Cosmos-2251 on 10 February 2009. In those days, there was not a uniform protocol for information exchange on space objects and debris to properly deal with an incident like that.

Once the issue was incorporated into the agenda of the STSC, a specific Working Group was established under the Chairmanship of Peter Martinez, national of South Africa,[10] that held sessions until 2018.

The United States provided a complement to the French proposal, suggesting dividing the topics into four clusters.[11] The Working Group established four expert groups for each area of work:[12] Expert Group A (on sustainable space utilization supporting sustainable development on Earth), Expert Group B (on space debris, space operations and tools to support collaborative space situational awareness), Expert Group C (space weather) and Expert Group D (on regulatory regimes and guidance for actors in the space arena). While these expert groups were a kind of deliberative body, the Working Group was instead the negotiating body.[13]

At that time, the Millennium Development Goals served as an excellent incentive for the work on sustainable development, viewed from the broader scope of the agenda of the United Nations. In particular, the Objective Seven (maintain environmental sustainability) was much connected to the goals pursued at COPUOS.

After 2015, a new UN global agenda with goals on sustainable development replaced the old driving force behind long-term sustainability at COPUOS. It was reinforced by the UNISPACE + 50 process, which enhanced the idea of space as a catalyst for socio-economic development. UNISPACE + 50 can be considered to be the most important event in COPUOS since the UNISPACE III Conference in 1999. Its leitmotiv was the commemoration of the 50th anniversary of the First Conference on Exploration and Peaceful Uses of Outer Space (UNISPACE I, in 1968).

As a result of the UNISPACE + 50 high-level segment, held on 20 and 21 June 2018, COPUOS agreed on a draft resolution that was adopted by the General Assembly without a vote. Its paragraph four encourages States to:

---

[10]Report of the 47th Session of the STSC (2010), UN Doc A/AC.105/958, paras. 181, 182.
[11]Long-term sustainability of outer space activities, UN Doc A/AC.105/C.1/2011/CRP.17, 7 February 2011.
[12]Terms of reference and methods of work of the Working Group on the Long-Term Sustainability of Outer Space Activities of the Scientific and Technical Subcommittee, UN Doc A/AC.105/C.1/L.307, 24 January 2011, para. 24; Nominations of members of expert groups and list of points of contact Communicated to the Secretariat as of 9 June 2011, UN Doc A/AC.105/2011/CRP.15 and Add.1, 9 June 2011.
[13]See Peter Martinez, "Space Sustainability", in *Handbook of Space Security. Policies, Applications and Programs*, ed. Kai Uwe Schrogl, Peter L. Hays, Jana Robinson, Denis Moura, Christina Giannopapa, (New York: Springer, 2015), 265.

continue to promote and actively contribute to strengthening international cooperation in the peaceful uses of outer space and the global governance of outer space activities, addressing challenges to humanity and sustainable development, ensuring the long-term sustainability of outer space activities and facilitating the realization of the 2030 Agenda for Sustainable Development, taking into account the particular needs of developing countries;[14]

While the most ambitious goal for COPUOS would have been to achieve consensus on the guidelines on long-term sustainability before UNISPACE + 50 to be able to refer them to the General Assembly during the same session, that could not be achieved even in a long negotiation at the 55th session of the STSC.

Despite that heated meeting in 2018, States achieved more flexible positions in 2019, which enabled the adoption of a first set of guidelines, fifty years after the first Moon landing.

3   The GRULAC Proposal on Long-Term Sustainability

During the 52nd session of the STSC in 2015, the GRULAC endorsed and made own a proposal originally submitted by Brazil in which the following elements may be distinguished: First, it proposed to include in the preamble a definition of sustainability built upon the outcome of the Conference on Sustainable Development Rio + 20, more precisely, the final document entitled 'The Future We Want'.[15]

It is relevant to recall that the concept of 'sustainable development' was coined in 1987 by the Brundtland Commission, created by the Board of the United Nations Environment Program to Develop an Environmental Perspective up to 2000.[16] The document entitled 'Our Common Future' referred to the needs of the present without compromising the ability to meet future needs.[17]

The definition that the GRULAC had originally proposed stated that sustainability was 'the need to adjust the objectives of access, exploration and use of outer space only for peaceful purposes with the need to preserve and protect the environment taking into account the needs of future generations'.[18] However, this wording did not reach consensus and it finally was drafted as 'the ability to maintain the conduct of space activities indefinitely into the future in a manner that

---

[14]Fiftieth anniversary of the first United Nations Conference on the Exploration and Peaceful Uses of Outer Space: space as a driver of sustainable development, UN Doc A/RES/73/6, 26 October 2018.

[15]The future we want, UN Doc A/RES/66/288, 27 July 2012.

[16]'Process of making environmental perspective to the year 2000 and beyond', UN Doc A/RES/38/161, 19 December 1983. Also known as the World Commission on Environment and Development was established by decision of the General Assembly A/RES/38/161.

[17]Report of the World Commission on Environment and Development, UN Doc A/42/427, 4 August 1987. Annex 'Our Common Future', paras. 27–30.

[18]Comments and proposed amendments to the Updated set of draft guidelines for the long-term sustainability of outer space activities (submitted by GRULAC), UN Doc A/AC.105/C.1/2015/CRP.19/Rev.1, 9 February 2015.

realizes the objectives of equitable access to the benefits of the exploration and use of outer space for peaceful purposes, in order to meet the needs of the present generations while preserving the outer space environment for future generations.'[19]

Second, the GRULAC proposal suggested including in guidelines that provided for consistency of national legislation with international space law the proviso that States review and amend legislation that contradicts such international legal standards. It also suggested adding language to the effect that States could not invoke national interest or national legislation to carry out actions contrary to the space governance regime.

Finally, the GRULAC proposed to include a new guideline that might be labelled as a 'non-proliferation clause', by which States should commit through their national legislation to developing space activities 'solely' for peaceful purposes. That guideline had to be complemented with an amendment in another one that should have included explicit language reaffirming the importance of preventing an arms race in outer space.

The first and third proposals had a common denominator, which was that outer space activities be preserved 'solely' for peaceful purposes. Unfortunately, that raised old discussions about how States interpret 'peaceful uses of outer space', and with that, arguments around the entitlement to use outer space for military non-aggressive purposes arose.[20]

The second part of the GRULAC proposal fared better, thus, review and amendment of national regulatory frameworks as needed was included in section II. A of the guidelines (dedicated to policy and regulatory framework for space activities).[21]

The third of the proposals was the most difficult to agree and, in fact, until the time of this publication States could not reach consensus on a non-proliferation clause (which for much of the negotiations was guideline 7). As already advanced, the proposal brought some complications around the term 'solely'. Concretely, delegations discussed whether to use 'only', 'exclusively'[22] or 'solely', and States exchanged views on whether any difference in meaning among them existed. In that

---

[19]Guidelines on Sustainability of Outer Space Activities (submitted by the Chair), UN Doc A/AC.105/C.1/L.366, 17 July 2018, para. 5.

[20]On this topic, see for instance, Pericles Gasparini Alves, "Prevention of an Arms Race in Outer Space. A Guide to the Discussions in the Conference of Disarmament", UNIDIR/91/79, 1991, Part I: 12; Steven Freeland, "The Laws of War in Outer Space", in *Handbook of Space Security. Policies, Applications and Programs*, ed. Kai Uwe Schrogl, Peter L. Hays, Jana Robinson, Denis Moura, Christina Giannopapa, (New York: Springer, 2015), 95.

[21]Report of the 62nd session of COPUOS (2019), UN Doc A/74/20, Annex II, guidelines A.1 and A.2.

[22]'Exclusively for peaceful purposes' is the wording used in art. IV of the Outer Space Treaty (1967) in its second paragraph, first sentence, which originated in a similar wording in art. 1 of the Antarctic Treaty (1959), which reads: 'Antarctica shall be used for peaceful purposes only'. Note, however, that art. IV of the Outer Space Treaty provides for the complete demilitarisation only of the moon and other celestial bodies, but not of outer space. This was then confirmed in art. 3 of the Moon Treaty (1979). Years later, this concept was taken up in art. 141 of the UN Convention on the Law of the Sea (1982), which provides that '(t)he Area shall be open to use exclusively for peaceful purposes by all States.'

regard, reference was made to the text of General Assembly Resolution 1348 (1958) that employed the word 'only'.[23] However, some delegations felt that such language had been overcome in successive documents and that the Outer Space Treaty did not include such terminology.

An inter-sessional meeting was convened between 5 and 9 September 2015 in which guideline 7 was again discussed, along with the inclusion of the non-weaponisation of outer space. The arguments against its inclusion were basically three: the absence of a definition of 'weapons', the lack of agreement on the issue at the Conference on Disarmament and the limitations imposed by the mandate of COPUOS.

The best result for the GRULAC would have been to include in this guideline both the non-weaponisation and a compromise to avoid an arms race, pointing at the report of the Group of Governmental Experts on Transparency and Confidence-building Measures in Outer Space Activities (A/68/189). However, this was not welcomed by other delegations; hence, various alternatives were examined, including reproducing the text of art. IV of the Outer Space Treaty, that provides for a clause of a limited arms control.[24]

In June 2016, the extended mandate of the Working Group was to terminate, but States had still not achieved any consensus on several guidelines (including guideline 7) or the preamble. Based on the mandate granted by COPUOS in 2015, the Chairman of the Working Group prepared a working paper for the 59th session of COPUOS in 2016. In that document, he broke down the guidelines into three categories: guidelines on which the Working Group was very close to achieving consensus; guidelines for which the Working Group might have reasonably expected to achieve consensus within the existing work plan (within which guideline 7 was) and, finally, guidelines for which the Working Group might have found it difficult to achieve consensus on all their constituent elements within the existing work plan.[25]

The last meetings of the Working Group concluded at the 53rd session of the STSC in 2016 with a stalemate difficult to overcome. On the one side, there were States that wished to submit the first set of guidelines to the consideration of the Committee and extend the mandate of the Working Group to address the remaining ones. On the other side, there were States that preferred to extend the mandate without adopting a first set. As States were not even able to reach a consensus on the final report of the Working Group, the future of the guidelines became uncertain, at least until the session of the Committee later that year.

Following the second inter-sessional meeting on 6 and 7 June, at the 59th session of the Committee in 2016, States achieved agreement on twelve guidelines (the so-called 'first set')[26] that would be ready for implementation by States and

---

[23]Question of Use of Outer Space, UN Doc RES 1348 (XIII), 13 December 1958, first preambular paragraph.

[24]See Steven Freeland, "The Laws of War in Outer Space", *supra* note 20, 95.

[25]Ideas for the way forward on the draft set of guidelines for the long-term sustainability of outer space activities, submitted by the Chair, UN Doc A/AC.105/C.1/2016/CRP.3, 28 January 2016.

[26]Report of the 59th Session of COPUOS (2016), UN Doc A/71/20, para. 130.

international organizations. At the same meeting, the Committee agreed to annex the first set of guidelines to the report of the session and extend the mandate of the Working Group for two additional years to work as a priority on the preamble and a second set of guidelines. Both sets of guidelines would form a compendium that would be referred in 2018 to the General Assembly (as already advanced, the year that coincided with the celebration of UNISPACE + 50).[27]

Between 19 and 23 September 2016, the third inter-sessional meeting took place, and a provisional definition on sustainability was achieved. However, no consensus could be reached on guideline 7; basically, there was no agreement about retaining the wording 'only for peaceful purposes'.

The fourth inter-sessional meeting was held in the margins of the Committee at its 60th session in 2017. The examination of the definition of sustainability continued, but the inclusion of the word 'solely' and the reference to the non-placement of weapons were firmly resisted. The same happened at the fifth inter-sessional meeting (2–6 October 2017), with the difference that this time the reference to art. IV of the Outer Space Treaty (incorporated in previous sessions) was also deleted.

After five inter-sessional meetings, the penultimate chance to negotiate the guidelines was the 55th session of the STSC in 2018. The possibility of including guideline 7 in the compendium to be referred to the General Assembly was at that stage very limited.

Another option would have been to include the content of this guideline in the report with which the compendium would be submitted to the Committee. However, in such a case, the contents of guideline 7 would not have been part of the compendium itself. The choice between taking a flexible position in order to submit to the Committee (at least) the partial results of many years of work and maintaining a tougher stance without any end-product became increasingly apparent. With that dilemma, no decision could be made on the mechanism for submission of the guidelines to the General Assembly, on how future work on implementation should be addressed, on mechanisms for review and incorporation of new guidelines and on the treatment of pending ones.

In the margins of the Committee at its 61st session in 2018, the Working Group continued in session, but to no avail. While some States were of the view that agreed upon and pending guidelines had to be annexed to the report of COPUOS for referral to the General Assembly, others were in favour of having a separate document with only agreed upon guidelines for referral to the General Assembly. A group of States (Australia, Canada, France, Germany, Israel, Italy, Japan, Netherlands New Zealand, United Kingdom and the United States) submitted a working document expressing their desire to have the agreed upon guidelines translated into all the UN official languages and referred them to the 73rd session of the General Assembly later that year for endorsement in a stand-alone resolution.[28]

---

[27]Ibid., para. 137.
[28]Proposal to adopt and refer to the General Assembly for endorsement the Compendium of Guidelines for the Long-Term Sustainability of Outer Space Activities (submitted by Australia, Canada, France, Germany, Israel, Italy, Japan, Netherlands, New Zealand, United Kingdom and United States), UN Doc A/AC.105/2018/CRP.26/Rev.2, 29 June 2018.

The Committee was unable to reach agreement on how to proceed. As a result, the work of the previous eight years remained in limbo: Neither would the agreed upon guidelines be referred to the General Assembly[29] nor was there agreement on the mechanism for further work on the seven remaining guidelines,[30] or on a mechanism for their implementation or review and incorporation of new ones.

The issue was taken up again at the meeting of the STSC in 2019, when positions were still divided on the need of such a mechanism and on further work on the guidelines. At the same meeting, and in a conciliatory attempt, Switzerland proposed organising a workshop at the beginning of the session of the Committee in June 2019 to exchange views on the future work and possible mechanisms for addressing the remaining guidelines. In a similar spirit, South Africa proposed submitting the 21 agreed guidelines to the General Assembly and establishing a mechanism to address the remaining ones, revise and implement them and incorporate new ones. That delegation also proposed that Brazil, as future Chair of the Committee, together with South Africa initiate informal consultations with interested delegations in order to be able to present a proposal for future work at the 62nd session of COPUOS in 2019.

That session in June was decisive for the future of the guidelines. The workshop was convened on the first day before the session of the Committee, when it became clear that it was necessary to work on long-term sustainability at COPUOS and its Subcommittees.[31] The delegations of Canada, France, Japan, United Kingdom and the United States submitted to COPUOS a working paper with a proposal to create a Working Group to implement the 21 agreed upon guidelines.[32]

United Arab Emirates submitted another proposal to establish a Working Group to continue the work on sustainability[33] and the Russian Federation, China, Nicaragua, Pakistan and Belarus submitted a joint proposal on the working modalities of a future Working Group on the subject.[34] The substantial difference between these two proposals was that the latter included explicitly in the mandate the treatment of the seven guidelines that had not reached consensus (among which is the old guideline 7).

---

[29]Guidelines on Long-Term Sustainability of Outer Space Activities (submitted by the Chair), UN Doc A/AC.105/C.1/L.366, 17 July 2018.

[30]Draft guidelines for long-term sustainability of activities in outer space (submitted by the Chair of the Working Group), UN Doc A/AC.105/C.1/L.367, 16 July 2018.

[31]Meeting hosted by Switzerland on possible further work on the long-term sustainability of outer space activities: Background and Chair's Summary, UN Doc A/AC.105/2019/CRP.16, 18 June 2018.

[32]Proposal for the establishment of a Working Group on implementation of agreed guidelines on long-term sustainability (submitted by Canada, France, Japan, United Kingdom and United States), UN Doc A/AC.105/2019/CRP.7/Rev.1, 19 June 2019, paras. 6–11.

[33]Proposal on the work related to Long-Term Sustainability of Outer Space Activities of the Committee on the Peaceful Uses of Outer Space (submitted by United Arab Emirates), UN Doc A/AC.105/2019/CRP.13, 13 June 2019.

[34]Proposal on the Modalities of the Working Group on the Long-Term Sustainability of Outer Space Activities of the Committee on the Peaceful Uses of Outer Space (submitted by Belarus, China, Nicaragua, Pakistan and the Russian Federation), UN Doc A/AC.105/2019/CRP.10/Rev.2, 20 June 2019.

The final decision of the Committee on the establishment of a Working Group struck a balance between all the proposals since the mandate encompasses implementation and incorporation of new guidelines taking into account existing documents, including the one containing the seven pending guidelines.[35] The STSC will decide on the composition of the bureau, the terms of reference and work plan at its next session in 2020. The Committee adopted the 21 guidelines and the preamble, which were annexed to the session report, and the General Assembly endorsed them in 2019.[36]

4 The Projected Code of Conduct of the European Union (EU)

Parallel to the discussion on long-term sustainability at COPUOS, the European Union, which was already an experienced space actor, had been working on the Code of Conduct for Space Activities (CoC) for several years. One of the differences between this effort and the sustainability process in COPUOS was that this initiative was top-down, meaning more political than technical.[37]

In 2002, following several failed attempts tò regulate arms control in outer space at the Conference on Disarmament, the Henry L. Stimson Center proposed a code of conduct containing a roadmap for the responsible use of outer space.[38] Years later, a similar idea would be collected and taken to a broader level by the European Union.

The origins of the CoC can be traced to 2006, when General Assembly Resolution 61/75 (2006) invited Member States to submit concrete proposals for maintaining international peace and security, promoting international cooperation and preventing an arms race in outer space.[39]

Later on, the General Assembly renewed its call to States to submit to the Secretary-General concrete proposals on transparency and confidence-building measures in outer space activities (TCBMs) through Resolution 62/43 (2007),[40] and two years later it requested that the Secretary-General submit a final report with an annex containing concrete proposals from States on TCBMs.[41]

---

[35]Report of the 62nd Session of COPUOS (2019), *supra* note 21, para. 167.
[36]Ibid, Annex II.
[37]See Peter Martinez, "Space Sustainability", *supra* note 13, 271.
[38]See Max M. Mutschler, "Security Cooperation in Space and International Relations", in *Handbook of Space Security. Policies, Applications and Programs*, ed. Kai Uwe Schrogl, Peter L. Hays, Jana Robinson, Denis Moura, Christina Giannopapa, (New York: Springer, 2015), 47.
[39]Transparency and Confidence-Building Measures in Outer Space Activities, UN Doc A/RES/61/75, 6 December 2006, para. 1.
[40]Transparency and Confidence-Building Measures in Outer Space Activities, UN Doc A/RES/62/43, 5 December 2007, para. 2.
[41]Transparency and Confidence-Building Measures in Outer Space Activities, UN Doc A/RES/64/49, 2 December 2009, para. 3.

In March 2007, Italy prepared a working paper entitled 'Food for thought for a Comprehensive Code of Conduct for Outer Space Activities'[42] and in September 2007, Portugal, on behalf of the European Union, provided information to the Secretary-General on the projected Code of Conduct on Space Objects and Space Activities, in response to the call made by General Assembly Resolution 61/75.[43]

It should be underscored that the European initiative was born and developed outside the multilateral framework of the United Nations. However, since 2008, the European Union began reporting on the projected CoC to COPUOS. Thus, France (with the rotating Presidency of the Council of the European Union in 2008) mentioned the initiative at the 51st session of COPUOS.[44] It was precisely that year that the Council of the European Union endorsed the first draft of the CoC.[45]

It was in this context that the negotiations on the issue of disarmament in outer space were deadlocked at the Conference on Disarmament. At that point, three initiatives on security in outer space coexisted, namely consultations on the establishment of a Working Group on Long-term Sustainability in COPUOS, a draft a treaty on non-weaponisation of outer space at the Conference on Disarmament, and the CoC.

Already then, some delegations had requested an analysis of the European initiative within COPUOS.[46] A year later, the draft text (already approved by the Council of the European Union in December 2008) was presented to COPUOS by the delegation of the Czech Republic on behalf of the European Union. On that occasion, the Committee noted in its report that the European Union was willing to carry out consultations and convene an ad-hoc international conference to sign the instrument.[47]

A second stage in the development of the European initiative began after the Council of the European Union gave a mandate to the High Representative for Foreign Affairs and Security Policy to conduct a series of consultations broadly with interested third States, in order to agree on an acceptable text for a larger number of States and subscribe it at a diplomatic conference.[48] At the 55th session of COPUOS in 2012, the promoters of the CoC reported on the draft CoC and advanced the idea of holding a diplomatic conference in 2013.

---

[42]Statement by the Alternate Representative and Charge d'Affaires ai of Italy to the United Nations, Ambassador Inigo Lambertini in multilateral negotiations on an 'international code of conduct on space activities' (27 July 2015). Available online at http://www.italyun.esteri.it/rappresentanza_onu/en/comunicazione/archivio-news/2015/07/2015-07-27-lambertini-spazio.html (accessed 28 August 2019).

[43]Report of the Secretary-General on Transparency and Confidence-Building Measures in Outer Space Activities (Portugal on behalf of the EU), UN Doc A/62/114/Add.1, 17 September 2007, paras. 3 and 4.

[44]Report of the 51st Session of COPUOS (2008), UN Doc A/63/20, para. 296.

[45]Council of the European Union, 'Council conclusions and draft Code of Conduct on space activities', 17175/08, 17 December 2008.

[46]UN Doc A/63/20, *supra* note 44, para. 301.

[47]Report of the 52nd Session of COPUOS (2009), UN Doc A/64/20, para. 45.

[48]Council of the European Union, "Conclusions of 27 September 2010 on a revised draft Code of Conduct on space activities", 14455/10, 11 October 2010.

Basically, the envisaged CoC established measures of transparency and confidence-building to safeguard the peaceful and sustainable use of outer space, preserving it for future generations. One of the clauses that raised criticism included 'the responsibility of States to refrain from the threat or use of force against the territorial integrity or political independence of any state, or in any manner inconsistent with the purposes of the Charter of the United Nations, and the inherent right of states to individual or collective self-defence as recognised in the Charter of the United Nations'.[49]

At that session of 2012, some States reported to COPUOS having participated in a meeting organised on 5 June 2012 in Vienna, where the initiative was assessed.[50] It was also in 2012 that US Secretary of State Hillary Clinton pledged its support for the European initiative, while making it clear that her country would not adhere to a code that limits the capacity to carry out activities in space or protect the United States or its allies.[51] For his part, Prime Minister of Australia Kevin Rudd supported the European initiative immediately thereafter.[52]

At the 56th session of COPUOS in 2013, the European Union expressed its firm resolve to develop a CoC in an open, transparent and inclusive manner. The European delegate also referred to a first round of consultations in Kyiv that had taken place on 16 and 17 May of that year, a second round of negotiations on 20 and 22 November that year in Bangkok and a third one would be on 27 and 28 May of the following year, in Luxembourg.[53]

Although the European Union tried at COPUOS to induce a larger number of States to adhere to the CoC initiative with a view to closing the process in an ad-hoc international conference,[54] concerns from some delegations became stronger. Criticism to the European initiative might be divided into two categories: procedural and substantive. As to the former, some delegations expressed a view that discussions on such matters should be carried out within the United Nations,[55] and more precisely within the overall context of the long-term sustainability in COPUOS.[56] Substantive criticism was particularly raised over the application of

---

[49]Version of 31 March 2014, available online at https://eeas.europa.eu/sites/eeas/files/space_code_conduct_draft_vers_31-march-2014_en.pdf (15 October 2019).

[50]Report of the 55th Session of COPUOS (2012), UN Doc A/67/20, para 46.

[51]United States Department of State press release, Hillary Rodham Clinton, Secretary of State, International Code of Conduct for space activities, 17 January 2012. Available online at http://www.state.gov/secretary/rm/2012/01/180969.htm (accessed 29 August 2019).

[52]"The International Code of Conduct on Outer Space Activities: An Australian Perspective", Space Sustainability Conference, Beijing, 8–9 November 2012. Available at: https://swfound.org/media/95008/kelly-australian_perspectives_icoc-nov2012.pdf (accessed 29 August 2019).

[53]Report of the 56th Session of COPUOS (2013), UN Doc A/68/20, para. 50.

[54]Council of the European Union, *supra* note 48.

[55]Report of the 57th Session of COPUOS (2014), UN Doc A/69/20, para. 55.

[56]Report of the 54th Session of the Legal Subcommittee (2015), UN Doc A/AC.105/1090, para. 201.

the concept of self-defence without having had a discussion in COPUOS on its application to outer space.[57]

However, the European Union continued working to finalise the CoC, to the point that in 2015 the Council of the European Union issued a decision supporting the initiative.[58] Objections became even stronger when the European Union announced during a COPUOS session that it was time to move from a consultative stage to a negotiating phase.[59]

The meeting organised by the European Union with the support of UNIDIR between 27 and 31 July 2015 indicated that it would not be possible to go ahead with the draft CoC in such a manner if the goal was to achieve a global initiative. Thus, the European Union admitted in the 59th session of COPUOS in 2016 that 'a non-legally binding agreement which is negotiated within the United Nations was the right way to proceed',[60] and from then on concentrated all its efforts and political commitment on advancing the process at COPUOS relating the compendium of guidelines on the long-term sustainability of space activities.

5 Other Initiatives Concerning the Security and Safety of Outer Space

Issues relating to security, safety and sustainability of outer space encompass an agenda that does not correspond to the mandate of a single body of the United Nations. Therefore, in parallel with the efforts in COPUOS to develop guidelines on long-term sustainability, the Conference on Disarmament and the First Committee of the General Assembly hosted discussions on non-weaponisation of outer space and on TCBMs, respectively.

In 2008, while the draft CoC was being prepared by the Working Group of the European Union on Global Disarmament and Arms Control to be later approved by the Council of the European Union, and Brachet presented its proposal on sustainability in COPUOS, the Russian Federation and China submitted to the Conference on Disarmament a first draft of a Treaty on the Prevention of Placement of

---

[57]Report of the 58th Session of COPUOS (2015), UN Doc A/70/20, para. 32; Russian assessment of the initiative and actions of the European Union to advance its draft code of conduct for outer space activities, UN Doc. A/AC.105/C.1/L.346, 30 July 2015.

[58]Council Decision (CFSP) 2015/203 of 9 February 2015 in support of the Union proposal for an international Code of Conduct for outer-space activities as a contribution to transparency and confidence-building measures in outer-space, OJ L 33/38, 10 February 2015.

[59]Achievement of a uniform interpretation of the right of self-defence in conformity with the United Nations Charter as applied to outer space as a factor in maintaining outer space a safe and conflict-free environment and promoting the long-term sustainability of outer space activities (submitted by the Russian Federation), UN Doc A/AC.105/C.1/2015/CRP.22, 2 February 2015, para. 10. On the position of the Russian Federation, see also, additional considerations and proposals to increase understanding of priorities, the overall meaning and functions of the concept and practice of ensuring long-term sustainability of activities in outer space (submitted by the Russian Federation), UN Doc A/AC.105/L.296, 30 April 2015, para. 5.

[60]Digital Recordings of the 59th Session of COPUOS (2016), 8 June 2016, 10 am (1:47:18).

Weapons in Outer Space, the Threat or Use of Force against Outer Space Objects (PPWT).

The idea was not to present a project on prevention of arms in outer space (PAROS), something on which the Conference on Disarmament had been working for almost three decades by then. Instead, the goal was to prohibit the weaponisation of outer space as a prior and necessary step to preventing an arms race (basically a preventive measure).[61]

That draft included a definition of 'weapon in outer space', which did not cover weapons on Earth that affect space objects in outer space.[62] It also included an article on the exercise of self-defence in accordance with art. 51 of the UN Charter (art. V) and one referring to the promotion of transparency and confidence-building measures (art. VI).[63]

The proposal contained two novel elements: the first one was the definition of outer space as the space above the Earth over 100 km above sea level[64] (the delimitation of air and outer space is an issue that COPUOS has not been able to settle in its deliberations). The second one was that the scope of the planned treaty did not cover only the weaponisation but also the use and threat of use of force. Thus, some authors have interpreted that the draft PPWT covered not only the placement but also the use of weapons, something absent in art. IV of the Outer Space Treaty.[65]

Despite these developments, the PPWT said nothing about testing, storage and development of such weapons. Neither did it provide a regime for compliance verification and monitoring.[66] Amid criticism and a general climate of stalemate in the Conference on Disarmament, the project did not reach consensus.

A new draft version was presented in June 2014[67] (just a couple of weeks after the third meeting organised by the European Union in Luxembourg for further discussion of the CoC), but the main problems of the previous version had not been

---

[61]Anton Vasiliev, "The Treaty on the Prevention of the Placement of Weapons in Outer Space, The Threat or Use of Force Against Outer Space Objects" in *Celebrating the Space Age: 50 Years of Space Technology, 40 Years of the Outer Space Treaty*, 2–3 April 2007, UNIDIR Conference Report, Geneva, Switzerland. Confront with the proposal contained in CD/1679, 28 June 2002.

[62]Peter L. Hays, "Developing Agile and Adaptive Space Transparency and Confidence-Building Measures", ESPI Report No. 27, September 2010, p. 32.

[63]Draft Treaty on Prevention of the Placement of Weapons in Outer Space and of the Threat or Use of Force Against Outer Space Objects, Conference on Disarmament, CD/1839 (29 February 2008). Available online at: http://www.un.org/ga/search/view_doc.asp?symbol=CD/1839 (accessed 29 August 2019).

[64]Art. 1(a) of the draft version of PPWT, 2008 version.

[65]See Steven Freeland, "The Laws of War in Outer Space", *supra* note 20, 104.

[66]Jana Robinson, "The Status and Future Evolution of Transparency and Confidence-Building Measures", *supra* note 1, 294.

[67]Letter dated 10 June 2014 addressed to the Acting Secretary-General of the Conference on Disarmament from the Permanent Representative of the Russian Federation and the Permanent Representative of China, for which the date in Chinese texts and Russian draft treaty are transmitted for the prevention of weaponisation of outer space and the threat or use of force against objects in outer space, presented by the Russian Federation and China, Doc. CD/ 1985, 12 June 2014.

overcome, such as the incorporation of Earth-based weapons and the lack of a verification mechanism.

That same year, at the request of the Russian Federation, the General Assembly adopted Resolution 69/32 'No first placement of weapons in outer space', with 126 votes in favour, four against[68] and 46 abstentions.[69] In that resolution, the General Assembly called upon States to start substantive work on the updated PPWT draft (operative paragraph no. three) and encouraged them to make a political commitment not to be the first to deploy weapons in outer space (operative paragraph no. five). While some authors have argued that support for this resolution could be interpreted as an indirect recognition of a possible acceptance of the PPWT,[70] the reality has not proven such a conclusion.

Some authors have pointed out that the unilateral declaration of not being the first to deploy weapons in space does not meet the conditions necessary to be considered a TCBM because it is not possible to demonstrate its implementation or verify its compliance.[71]

Between the two PPWT draft versions, in 2011 the General Assembly adopted Resolution 65/68 with 167 votes in favour and one abstention (United States), requesting that the Secretary-General establish a Group of Governmental Experts (GGE) with geographical representation to produce a report on TCBMs in outer space, and decided to include a tentative agenda item on that matter.[72]

TCBMs can be defined as governmental measures to exchange information in order to reduce misperceptions and miscalculations, thus preventing confrontations and military escalations and promoting international stability. Such measures can take various forms and names, such as good practices, codes of conduct, guidelines or rules of conduct.[73]

The GGE consisted of experts from 15 countries[74] (many of whom were national delegates to COPUOS) that met three times (one in 2012 and two in 2013). In 2013, it produced a report with recommendations on TCBMs to promote the 3S, including the exchange of information on space, on international cooperation, notification and information on launches, space debris, potential collisions and on other hazards to

---

[68]United States, Israel, Georgia, Ukraine.

[69]Mostly countries of the European Union, Australia and Canada.

[70]Hao Liu, Fabio Tronchetti, "United Nations Resolution 69/32 on the "Do not first placement of weapons in space": A step forward in the prevention of an arms race in outer space", *Space Policy* (2016), 4.

[71]See Peter Martinez, Richard Crowther, Sergio Marchisio, Gerard Brachet, "Criteria for Developing and Testing Transparency and Confidence-Building Measures (TCBMs) for Outer Space Activities", *Space Policy* (2014), 2.

[72]Measures Transparency and Confidence-Building in Outer Space, UN Doc A/RES/65/68, 8 December 2010.

[73]Jana Robinson, "The Status and Future Evolution of Transparency and Confidence-Building Measures", *supra* note 1,294.

[74]Brazil, Chile, China, France (Gerard Brachet, former President of COPUOS and promoter of the inclusion of sustainability on the agenda of COPUOS), Italy (Sergio Marchisio, former Chairman of the Expert Group D on space governance), Kazakhstan, Korea, Russian Federation, Nigeria, Romania, South Africa (Peter Martinez, former Chairman of the Working Group on Sustainability), Sri Lanka, Ukraine, United Kingdom and United States.

space objects.[75] In its conclusions, the GGE supported efforts to achieve political commitments to encourage responsible behaviour in outer space and, among the examples cited, mentioned 'a code of conduct' (possibly referring to the European initiative).[76] That same year, the General Assembly welcomed the report of the GGE and called upon States to review and implement the recommended TCBMs.[77]

This chapter cannot conclude this section without referring to the joint work between the First and Fourth Committee of the General Assembly. Given the interconnectedness between safety, security and sustainability of outer space activities, and being addressed in various UN bodies, the General Assembly decided to convene a joint meeting of the First Committee (Disarmament and International Security) and the Fourth Committee (Special Political and Decolonization) within the framework of the endeavours on TCBMs.[78]

In that event, the Director of the Office for Outer Space Affairs, Simonetta di Pippo, made it clear that there is a complex and evolving agenda in the field of space affairs and sustainability, specifically including long-term sustainability of space activities.[79]

After that first milestone of 2015, the General Assembly decided to convene a joint round table of the First and the Fourth Committee in 2017, as a contribution to the 50th anniversary of the Outer Space Treaty.[80] On that occasion, several challenges to security, safety and sustainability of activities in outer space were addressed, such as space debris, near-Earth objects, the emergence of new space actors, the development of anti-ballistic missiles, the practice of surveillance satellites and the extraction of natural resources in space.[81] In addition, some States recognised that the right to self-defence enshrined in art. 51 of the UN Charter applied to the context of outer space and a delegation concluded that the resort to that right required further study in the context of outer space.[82]

---

[75]Note by the Secretary-General containing the report of the Governmental Expert Group on Transparency and Confidence-Building Measures in Outer Space Activities, UN Doc. A/68/189, 29 July 2013.
[76]Ibid., para. 69.
[77]UN Doc A/RES/65/68, *supra* note 72.
[78]Transparency and Confidence-Building Measures in Outer Space Activities, UN Doc A/RES/69/38, 2 December 2014, para. 6.
[79]See Introductory remarks by the Director of OOSA, October 22, 2019. Available online at http://www.unoosa.org/oosa/en/aboutus/director/director-statements/director-speech-2015-joint-ad-hoc-meeting-ga-1st-4th-committee.html (accessed 29 August 2019).
[80]International Cooperation in the Use of Outer Space for Peaceful Purposes, UN Doc A/RES/71/90, 6 December 2016, para. 15.
[81]Joint Panel Discussion of the First and Fourth Committees on possible Challenges to Space Security and Sustainability, para. 13. Available online at http://www.unoosa.org/documents/pdf/gajointpanel/Co-Chair_Summary_C1-C4_Joint_Panel_Discussion_Final_2.pdf (accessed 29 August 2019).
[82]Ibid., para. 1.5.

Based on the positive experience of 2017, a year later COPUOS recommended that a new joint panel of the First and the Fourth Committee take place in October 2019,[83] which suggests that the trend for the future may be to institutionalise this practice.

The mandate of this meeting had four indicative themes: (1) identify intersections between security and sustainability, (2) take stock of the ongoing processes on the matter, (3) exchange views on international and (4) cooperation identify approaches that help achieve targets for safety and sustainability.[84]

6 Conclusions on the Future of Sustainability: The Road Ahead

While the projected CoC did not materialise, it is important to note that the set of guidelines that COPUOS adopted foresees a series of commitments that were also included in that initiative, such as the need to adopt measures and policies to reduce the risks of collision, interference and creation of space debris; the need to share information regarding launches into space and space events, as well as those related to space weather (Part II.B of the Guidelines; Part III of the draft CoC).

International cooperation is also guaranteed in an equitable and mutually acceptable basis, according to the 'Declaration on International Cooperation in the Exploration and Use of Outer Space for the benefit and interest of all States, taking particular account of the needs of Developing countries'[85] (Part I, paragraph 19 of the Guidelines; Part I, paragraph 42 of the draft CoC).

While both the Guidelines (Part I, paragraph 22) and the projected CoC (Part III, Section 7) refer to art. IX of the Outer Space Treaty and encourage resolving issues with States involved, they differ in that the Guidelines state that the outcome of such consultations should be presented to COPUOS if the parties involved consent.

As for the GRULAC proposal, the balance is also positive since the preamble that was adopted includes a definition of sustainability and provides for the review and amendment of national space legislation that is contrary to the international governance regime. However, the result is not entirely as expected, since the reference to the use of outer space 'only' or 'exclusively' for peaceful purposes did not reach consensus in the preamble, and it is uncertain if the original guideline 7 or a similar one will be adopted in the future. At the current stage, States do not seem to

---

[83]Report of the 61st Session of COPUOS (2018), UN Doc A/73/20, para. 385. The General Assembly decided to convene the joint panel through resolution A/RES/73/91, 7 December 2018.
[84]Draft Concept note on the joint panel discussion of the First and Fourth Committees of the General Assembly on possible challenges to space security and sustainability, UN Doc A/AC.105/2019/CRP.19, 21 June 2019.
[85]'Declaration on International Cooperation in the Exploration and Use of Outer Space for the benefit and Interest of all States, Taking Particular Account of the Needs of Developing Countries', UN Doc A/RES/51/122, 4 February 1997.

be ready to build consensus on a clause that goes beyond the terms of art. IV of the Outer Space Treaty, to provide for a total ban on the weaponisation of outer space.

Clearly, there are specific points that at the time of this publication present difficulties to achieving agreement: the need for a binding instrument to prevent the placement and use of all types of weapons in outer space, the need to discuss at COPUOS the exercise of self-defence, a common understanding of 'space weapons' and new threats to space activities, such as cyber and electromagnetic attacks.

The comprehensive analysis made in this article of various initiatives has attempted to demonstrate that subjects concerning 3S are interconnected, despite its fragmented treatment in various UN bodies. The underlying idea is to convey that it is not possible to conceive the long-term sustainability of space activities without a commitment to ensuring safety and security of outer space activities.

Therefore, the major challenge that COPUOS faces is to continue addressing future work on sustainability without waiting for progress to be made in other fora. To this end, it is necessary to envisage the work on sustainability as a renewed effort to overcome deadlocks that prevent progress on safety and security of outer space. Ultimately, the approach proposed here is to leverage the work of COPUOS on sustainability as a facilitator of dialogue to achieve consensus in other bodies, and to not use the failures of other bodies to block the work at COPUOS.

Laura Jamschon Mac Garry holds a Law Degree (University of Buenos Aires), a LLM (University of Vienna, Faculty of Legal Sciences) and is a PhD candidate at the Sapienza University of Rome, Department of Political Science, International Law and Human Rights. Her main areas of research are international law and multilateral discussions on space matters and cybersecurity.

Latin America's Space-Related Heritage and Its Preservation

Annette Froehlich⊙ and André Siebrits⊙

Abstract

Latin America has a longstanding relationship with space. This chapter explores this relationship around the theme of heritage. It begins by characterising this relationship as both 'bottom-up' and 'top-down', and then analyses the theme of heritage in the context of pre-Columbian cultural astronomy (and related world heritage sites) as well as modern astronomy, followed by a review of modern regional collaboration in relation to Earth observation, particularly in terms of monitoring and protecting world heritage sites. It then considers the importance of protecting access to the dark night sky (threatened around the world by light pollution) as an important facet of heritage (and modern astronomy). Finally, it considers some of the concerns about satellite mega-constellations and their potential impact on this heritage for future generations.

1 Introduction

In terms of the relationship of Latin America with space, for the purposes of this chapter a 'bottom-up' and a 'top-down' approach will be analysed in relation to heritage. This emphasises, on the one side, Latin America's strong historical and modern astronomical traditions and heritage ('bottom-up') and, on the other, its contemporary role as a region that is actively participating in Earth observation, and

A. Froehlich
SpaceLab, University of Cape Town, Rondebosch, South Africa
e-mail: Annette.Froehlich@uct.ac.za

A. Siebrits (✉)
Department of Political Studies, University of Cape Town, Rondebosch, South Africa
e-mail: SBRAND003@myuct.ac.za

© Springer Nature Switzerland AG 2020
A. Froehlich (ed.), *Space Fostering Latin American Societies*, Southern Space
Studies, https://doi.org/10.1007/978-3-030-38912-3_11

advancing its spacefaring capabilities and heritage ('top-down'). This Earth observation capability also extends to the monitoring and protection of world heritage sites (including those in Latin America), many of which have a strong relationship with astronomical observation, thus creating the circular model of Latin America's relationship between space and heritage (Fig. 1).

2 Latin America's Affinity with the Night Sky

This section begins with Latin America's Earth-bound astronomical observations, both historical and modern.

2.1 Pre-Colombian Astronomical Heritage

Cultural astronomy can be defined as "the use of astronomical knowledge, beliefs or theories to inspire, inform or influence social forms and ideologies, or any aspect of human behaviour".[1] Ethnoastronomy and archaeoastronomy both fall under cultural astronomy. Ethnoastronomy can be understood as "the branch of astronomy that involves learning about the astronomical system of non-Western people"[2] while archaeoastronomy is "the study of how people have understood, conceptualized and used the phenomena in the sky and what role the sky played in their cultures, by analysing their material remains".[3] Latin America is rich in archeoastronomical sites, which in turn reveal much about the ethnoastronomical heritage of pre-Columbian times. The United Nations Educational, Scientific and Cultural Organisation (UNESCO), in collaboration with the International Astronomical Union (IAU) have recognised the value of these sites, and they declare that:

> Heritage that bears witness to people's interpretation and understanding of the sky from earliest times through to the present day stands as a record of the extraordinary diversity of ways in which our species has viewed, interpreted and understood the relationship between itself and the world—the universe—that we inhabit. If we are to keep this record intact, it is vital to recognize and safeguard cultural sites and natural landscapes that encapsulate and epitomize the connection between humankind and the sky.[4]

---

[1]Campion quoted in J. O. Urama, "Cultural Astronomy in Nigeria", *African Skies/Cieux Africains 15*, (2011): 7.
[2]Jhan-Curt Fernandez, "Ethnoastronomy: The People and the Stars", 26 August 2016, https://www.slideshare.net/jcurtfernandez/ethnoastronomy-the-people-and-the-stars (accessed 12 November 2019).
[3]Fábio Silva, "Who is an archaeoastronomer?", *Space Awareness*, no date, http://www.space-awareness.org/bg/careers/career/who-archaeoastronomer/ (accessed 12 November 2019).
[4]UNESCO and IAU, "Why preserve astronomical heritage?", *Portal to the Heritage of Astronomy*, no date, https://www3.astronomicalheritage.net/index.php/why-preserve-it (accessed 12 November 2019).

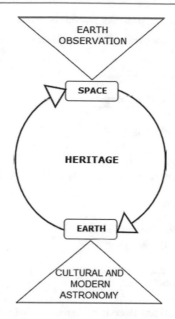

Fig. 1 The circular model of Latin America's relationship between space and heritage

Jointly, UNESCO and the IAU have launched an initiative dedicated to astronomical heritage, and they maintain the Portal to the Heritage of Astronomy, in which they define heritage as such:

> Astronomical heritage does not only come in the form of fixed, tangible heritage (sites and landscapes). It is equally important to recognize and protect movable objects and artefacts, the intangible heritage of astronomy in various forms, and natural heritage relating to astronomy—including the visibility of the dark night sky itself.[5]

Numerous UNESCO World Heritage Sites in Latin America have a connection to astronomy or indigenous astronomy. Of the 79 World Heritage Sites thus identified, 23 are located in Latin America, with the majority of these located in Mexico and Peru. The sites that have such a connection to indigenous or cultural astronomy are outlined in Table 1.

Four other pre-Colombian sites have also been identified as having links to astronomy, as outlined in Table 2.

---

[5]Ibid.

Table 1 Established world heritage sites with possible connections to astronomy

• Tiwanaku: Spiritual and Political Centre of the Tiwanaku Culture • Fuerte de Samaipata	Bolivia
• Rapa Nui National Park	Chile
• San Agustín Archeological Park • National Archeological Park of Tierradentro	Colombia
• Tikal National Park • Archaeological Park and Ruins of Quirigua	Guatemala
• Maya Site of Copan	Honduras
• Rock Paintings of the Sierra de San Francisco • Historic Centre of Oaxaca and Archaeological Site of Monte Albán • Pre-Hispanic Town of Uxmal • Pre-Hispanic City of Teotihuacan • Pre-Hispanic City of Chichen-Itza • Pre-Hispanic City and National Park of Palenque • Historic Centre of Mexico City and Xochimilco • El Tajin, Pre-Hispanic City • Archaeological Monuments Zone of Xochicalco • Ancient Maya City of Calakmul, Campeche	Mexico
• Río Abiseo National Park • Lines and Geoglyphs of Nasca and Pampas de Jumana • Historic Sanctuary of Machu Picchu • City of Cuzco • Chavin (Archaeological Site)	Peru

UNESCO and IAU, "Theme: Indigenous Uses of Astronomy," *Portal to the Heritage of Astronomy*, no date, https://www3.astronomicalheritage.net/index.php/show-theme?idtheme=9 (accessed 12 November 2019)

Table 2 Cultural tangible fixed astronomical heritage sites

• Boca de Potrerillos	Mexico
• Caguana	Puerto Rico (USA)
• Chankillo	Peru
• Viña del Cerro	Chile

UNESCO and IAU, "Astronomical Heritage Finder", *Portal to the Heritage of Astronomy*, no date, https://www3.astronomicalheritage.net/index.php/heritage/astronomical-heritage-finder (accessed 17 November 2019)

Latin America is thus rich in pre-Colombian astronomical heritage, which is accessible today due to the historical accounts of Spanish chroniclers, as well as through "reports by ethnographers and anthropologists working among native groups during the last 150 years, [that have identified] several well-presented archaeological and historical sites, numerous rock-art sites and the living traditions of indigenous peoples".[6] For the purposes of this chapter, the region of Latin America can be split into two—Mesoamerica and South America.

Astronomical observations played a key role in both these regions, for example in timekeeping, which was derived from "observations of the recurrent phenomena perceived in the sky", and calendars, which were used to observe the passage of time within each year, not to mark the time from any date in particular.[7] Moreover, it is noteworthy that most Native American populations used a lunar calendar, and "the observation of the phases of the moon might well have been instrumental in shifting attention from irregular time indicators to continuous time reckoning".[8] This was not the only instance of astronomical observations being used for time-keeping, as observations of planets or stars also played a role, since the "(heliacal) rising of specific constellations or asterisms (most commonly the Pleiades or the stars of Orion) served to mark the beginning of the year".[9]

The impact of astronomy on Native American societies beyond timekeeping was also significant in the context of cultural astronomy. The knowledge, beliefs, and theories derived from astronomical observations had a physical manifestation in daily life:

> The creation of order is a feature common to all Native American societies and its physical manifestations may be perceived in a wide range of material remains left by them. A common characteristic is that actions such as centring settlements, ordering dwellings and aligning burials stem from a shared knowledge of origins ('origin myths') and communal world-views that embody a variety of astronomical and cosmogonic principles. The ability to design and orient dwellings, shrines, temples, palaces or burials according to the annual movements of the sun is testimony to the knowledge of the sky possessed by many Native American groups.[10]

2.2   Modern Astronomical Heritage

UNESCO and the IAU also maintain a database of contemporary astronomical heritage sites that are "outstanding in the history of astronomy but do not necessarily demonstrate potential Outstanding Universal Value which would be needed for inscription on the World Heritage List". One of these is located in Latin America, namely "the La Plata Astronomical Observatory" (Observatorio Astronómico de La Plata), Argentina.[11] This Astronomical Observatory was one of the first institutions of its kind established—"[t]he idea of erecting an observatory was

---

[6]UNESCO and IAU, "Theme: Pre-Columbian America", *Portal to the Heritage of Astronomy*, no date, https://www3.astronomicalheritage.net/index.php/show-theme?idtheme=8 (accessed 17 November 2019).

[7]Ibid.

[8]Ibid.

[9]Ibid.

[10]Ibid.

[11]UNESCO and IAU, "IAU - Outstanding Astronomical Heritage", *Portal to the Heritage of Astronomy*, no date, https://www3.astronomicalheritage.net/index.php/heritage/outstanding-astronomical-heritage (accessed 16 November 2019).

inspired by the event of the transit of Venus, which was observed by French and Argentinian astronomers together".[12] Since its founding it has "became one of the leading institutions of the southern hemisphere".[13] Another modern astronomical heritage site, located in Chile, is the AURA Observatory, with mountaintop groups of telescopes on Cerro Tololo and Cerro Pachón.[14] This observatory is one of the leading astronomical institutions in the world, and work done here has been awarded the Nobel Prize for Physics.[15]

3 Contemporary Latin American Collaboration and Projects

The sky that attracts great interest in Latin America is also nowadays very inspiring for studies and regional collaboration projects in relation to Earth observation and protecting heritage sites.

Latin American space cooperation projects, such as the SABIA (SABIA—Satélite Argentino-Brasileño de Información en Alimento, Agua y Ambiente) satellite programme, are increasing. SABIA is an Argentine-Brazilian satellite for information on food, water and the environment that goes by the name of SABIA-Mar (SAC-E), which is part of the SAC series. This Earth observation satellite is scheduled to be launched in 2022.[16] The Argentinean Commission for Space Activities CONAE (Comisión Nacional de Actividades Espaciales), the Brazilian Space Agency (AEB), and the Brazilian Instituto Nacional de Pesquisas Espaciais (INPE) are conducting this Earth observing mission within the framework of a space cooperation program. The objective of this mission is the observation and monitoring of the oceans, ecosystems, carbon cycles, and coastal and marine habitats of both countries.[17] SABIA-Mar 1 is to be built by Argentina, while SABIA-Mar 2 will be built by Brazil.[18]

---

[12]UNESCO and IAU, "Category of Astronomical Heritage: Tangible Immovable, La Plata Astronomical Observatory, Argentina", *Portal to the Heritage of Astronomy*, no date, https://www3.astronomicalheritage.net/index.php/show-entity?identity=122&idsubentity=1 (accessed 16 November 2019).

[13]Ibid.

[14]UNESCO and IAU, "Category of Astronomical Heritage: Cultural-Natural Mixed Windows to the Universe (Multiple Locations): AURA Observatory, Chile", *Portal to the Heritage of Astronomy*, no date, https://www3.astronomicalheritage.net/index.php/show-entity?identity=59&idsubentity=5 (accessed 16 November 2019).

[15]Ibid.

[16]Gunter's Space Page, SABIA-Mar 1 (SAC E), https://space.skyrocket.de/doc_sdat/sabia-mar-1.htm (accessed 10 November 2019).

[17]Statement by Argentina at the 50th session of the Scientific and Technical Subcommittee of UNCOPUOS, 18.2.2013, Vienna, p. 2, pt. c; Statement by Brazil at the 50th session of the Scientific and Technical Subcommittee of UNCOPUOS, 13.2.2013, p. 6.

[18]Gunter Dirk Krebs, "SABIA-Mar 2", Gunter's Space Page, 12 May 2018, https://space.skyrocket.de/doc_sdat/sabia-mar-2.htm (accessed 4 January 2020).

Another joint project in this region is the monitoring of world heritage sites via satellite data. Several projects have been launched in cooperation with UNESCO.[19] For example, Argentina, Brazil and Paraguay have initiated the Iguazú World Heritage[20] and the Galapagos World Heritage Site of Ecuador is monitored with the support of the Argentinean CONAE.[21]

Of particular interest was Peru's initiative in 2002 to propose the Andean Main Road as a World Heritage Site, which other concerned countries subsequently joined.[22] The Qhapac Ñan or Camino del Inca was the main Andean road that served as the backbone of the Inca Empire.[23] Argentina, Bolivia, Chile, Colombia, Ecuador and Peru have brought together their efforts to detect and monitor the old Inca path, and the impact of human activities over this unique archaeological heritage.[24] This is a critical initiative because "Today, the cultural landscapes of *Qhapaq Ñan* form an exceptional backdrop on which living Andean cultures continue to convey a universal message: the human ability to turn one of the harshest geographical contexts of the American continent into an environment for life".[25] Indeed, this main Andean road connected the Inca empire from the peaks of the Andes, with an altitude of more than 6.000 m, in a north-south direction along the Pacific coast. Towns, centres of production and of worship were linked together in this Inca political project. In this context it is noteworthy that Argentina organised a workshop in 2012 on the Use of Space Technology for the Conservation of World Heritage Sites for Latin American and Caribbean Experts that was attended by conservation experts throughout the region.[26]

---

[19]European Space Agency and UNESCO 'Open Initiative on the use of space technologies to support the World Heritage Convention' for further information: http://www.unesco.org/science/remotesensing/?id_page=86&lang=en (accessed 10 November 2019).

[20]Iguazú National Park was created to ensure the conservation of the Iguazú falls and the biodiversity in its surroundings. Remote sensing data are useful for site monitoring including difficult access areas, detection of activities on the site (e.g. agriculture or human movements), and wildlife populations.

[21]CONAE, Space Activities in Argentina and World Heritage Sites Conservation, Félix Menicocci (CONAE), 28.09.2012, Naples.

[22]IGN, QHAPAQ ÑAN: Patrimonio Mundial de la Humanidad, http://www.ign.gob.ar/content/qhapaq-%C3%B1-patrimonio-mundial-de-la-humanidad (accessed 10 November 2019).

[23]UNESCO, Qhapaq Ñan, Andean Road System, https://whc.unesco.org/en/list/1459/ (accessed 10 November 2019).

[24]CONAE, Space Activities in Argentina and World Heritage Sites Conservation, Félix Menicocci (CONAE), 28.09.2012, Naples.

[25]UNESCO, Qhapaq Ñan, or Main Andean Road, https://whc.unesco.org/en/qhapaqnan/ (accessed 10 November 2019).

[26]CONAE, Space Activities in Argentina and World Heritage Sites Conservation, Félix Menicocci (CONAE), 28.09.2012, Naples.

4   Protection of the Dark Night Sky as Our Common Cultural Resource

The observation of the starry sky is our common cultural resource and the astronomical world heritage dating from early humanity. However, the observation and benefit of this heritage is becoming threatened, and an undisturbed view to the sky firmament is no longer possible from various places on Earth.

Consequently, the World Heritage Center of UNESCO, together with the IAU, has launched the Internet portal "Portal to the Heritage of Astronomy"[27] to emphasise the importance of the astronomical world heritage and its protection and preservation for the benefit of the whole of humanity. Indeed, the world of science is connected with the cultural sphere, meaning humanity is connected with heaven. The starry sky is our common and universal world heritage. Since time immemorial and in all parts of the world, astronomical institutions have been the most complex and diverse way in which people have perceived and tried to understand the cosmos. The astronomical world heritage list includes sites such as the Strasbourg Cathedral with its astronomical clock (F), the Royal Observatory in Greenwich (GB), the Einstein Tower in Potsdam (D), Stonehenge (GB), and the aforementioned La Plata Astronomical Observatory in Argentina.

Nevertheless, observatories are being threatened and an undisturbed view of the sky firmament is sometimes no longer possible. Therefore, reference should be made to the Declaration in Defence of the Night Sky and the Right to Starlight (La Palma Declaration),[28] which aims to protect the unrestricted view to the starry sky as part of the "Starlight Initiative".[29] It considers the starlight as the common heritage of the whole of humanity, as the starry sky has been the source of inspiration for humanity since time immemorial. Observation of the starry sky has had significant influence on the development of all cultures and civilisations over the centuries. Contemplation of the firmament has inspired many scientific and technical developments that have contributed to human development. Indeed, humankind has always watched the sky, either to interpret it or to detect physical laws governing the universe.

However, the Declaration of La Palma also notes that the quality of the night sky and thus the accessibility to starlight have deteriorated in an alarming manner, so that the observation of the starry sky is increasingly difficult. This leads to a loss of cultural and scientific resources with unpredictable consequences. According to this Declaration, pertinent standards for preserving the quality of a clear and unpolluted night sky should be adopted by all states to guarantee the common right to contemplate the firmament. This should be seen as an indispensable human right, equal to all other environmental, social and cultural rights, as these affect the development of all people and the preservation of biodiversity (point 1). The increasing

---

[27]UNESCO, Portal to the Heritage of Astronomy, https://www3.astronomicalheritage.net/.
[28]Declaration in Defence of the Night Sky and the Right to Starlight (La Palma Declaration), Canary Islands, 19–20 April 2007.
[29]ISSUU, Starlight Initiative, https://issuu.com/starlightinitiative (accessed 10 November 2019).

deterioration of the night sky must therefore be regarded as a threat like resource exploitation or pollution (point 2). The protection and preservation of the natural and cultural heritage in relation to the nocturnal skyscape and its observing opportunities is both a fundamental right and a cross-border obligation to preserve the quality of life.

Furthermore, the Declaration of La Palma states that the opportunity to discover the nocturnal starry sky is an important part of humanity's heritage (point 3) as it generates knowledge in our society. The harmful effects of emissions and the increasing influence of artificial light on the nocturnal sky atmosphere even in protected areas have serious effects on living beings and ecosystems. Control of light emissions should therefore be the cornerstone of the protection of our nature and biodiversity (point 5), as certain living beings need darkness to survive (in contrast to humans, who in some cities have turned night to day to increase productivity). As a result, people living in megacities may have never seen the third dimension of the starry sky in their lives, caught between the urban canyons of the big cities and thus deprived of the heritage to position themselves as human beings in the universe. Moreover, the intelligent use of artificial light, which reduces the glow of the sky, leads also to a more efficient use of energy (point 7). In fact, the amount of uselessly irradiated light has risen dramatically in recent decades. Consequently, places where unrestricted observation of the starry sky is possible should be specially protected and preserved, i.e. as a small compensation for what they have contributed to our knowledge, science, and technological developments (point 8) in the past, and to protect the possibilities of future astronomical discoveries.

Fortunately, some legislative initiatives have been undertaken to limit the influence of light on certain astronomical places, such as in Spain with the Ley del Cielo (1988)[30] and in Chile with the Norma de la Contaminación Lumínica (1999), to keep the country as a coveted astronomical location for world-renowned observatories such as the European Southern Observatory (ESO) which operates telescopes in the Atacama Desert of Chile (La Silla Observatory, Paranal Observatory).

5 Conclusion: Reserving Space-Related Heritage in an Era of Satellite Mega-Constellations

Despite the urgent need to protect unrestricted views and observations of the starry night sky, new threats are arising, including from potential mega-satellite constellations. Several companies have announced plans to orbit thousands of smaller satellites in the next few years. They claim that such constellations will benefit humankind by providing the Internet to populations in wide areas in Latin America

---

[30]Ley del Cielo (Ley 31/1988) LEY 31/1988 de 31 de octubre, sobre Protección de la Calidad Astronómica de los Observatorios del Instituto de Astrofísica de Canarias.

and Africa. However, the field of astronomy, and the clear views it requires (for instance in contemplating star constellations) will be immeasurably affected if thousands of satellites permanently orbit though those constellations. It is clear that despite the promise of these new satellites, "We do not yet understand the impact of thousands of these visible satellites scattered across the night sky and despite their good intentions, these satellite constellations may threaten both [advancing our understanding of the Universe and protection of nocturnal wildlife]".[31] Therefore, more efforts should be undertaken to preserve the space-related heritage by encouraging all countries to take part in the technical and social developments in this arena, which will be a challenge for the coming years. Humanity has, as this chapter argued, had a long and fruitful relationship with the night sky, including in Latin America, and while striving to advance in the field of satellite mega-constellations, it is of the utmost importance to preserve to the greatest degree possible clear and unobstructed access to the night sky. This is the heritage that was handed down from past generations, and which we must preserve for generations to come. Only by including all possible stakeholders in discussions, and by increasing public awareness of the importance of the night sky, can we hope to achieve the required balance between progress and preservation. These themes of heritage and the night sky will be explored to a greater extent in future research.

Dr. Annette Froehlich is a scientific expert seconded from the German Aerospace Center (DLR) to the European Space Policy Institute (Vienna), and Honorary Adjunct Senior Lecturer at the University of Cape Town (SA) at SpaceLab. She graduated in European and International Law at the University of Strasbourg (France), followed by business oriented postgraduate studies and her PhD at the University of Vienna (Austria). Responsible for DLR and German representation to the United Nations and International Organizations, Dr. Froehlich was also a member/alternate head of delegation of the German delegation of UNCOPUOS. Moreover, Dr. Annette Froehlich is author of a multitude of specialist publications and serves as a lecturer at various universities worldwide in space policy, law and society aspects. Her main areas of scientific interest are European Space Policy, International and Regional Space Law, Emerging Space Countries, Space Security and Space & Culture. She has also launched as editor the new scientific series "Southern Space Studies" (Springer Publishing House) dedicated to Latin America and Africa.

André Siebrits is a South African researcher focusing on the space arena (especially in developing world contexts), education and the use of educational technologies, and International Relations (particularly in the Global South). He is currently working with the European Space Policy Institute (Vienna), and has experience as an e-learning researcher and as an African political risk analyst. He graduated with a Master of Arts in International Studies from the University of Stellenbosch, where his research revolved around theories of International Relations. He is currently a PhD Candidate at the Department of Political Studies at the University of Cape Town (UCT), where his research focuses on the role of the Global South in the space arena, especially in relation to governance, seen from an International Relations perspective. André is an author of publications in the e-learning field, and has written on the space-education ecosystem for

---

[31]A. J. Mackenzie, Who speaks for the night sky? The Space Review, 10 June 2019, http://www. thespacereview.com/article/3728/1 (accessed 10 November 2019).

sustainability and the role of educational technologies in Africa, on integrated space for African society (legal and policy implementation of space in African countries—specifically Algeria, Morocco, Tunisia, and Zimbabwe), and on the African space arena. André has also presented lectures at the UCT SpaceLab (for their Space and Society course) on the African space arena and on the role of educational technologies in space education in Africa.

Printed in the United States
by Baker & Taylor Publisher Services